W9-BXX-776

WHAT JUST HAPPENED

WHAT JUST HAPPENED

A CHRONICLE FROM THE INFORMATION FRONTIER

James Gleick

Pantheon Books, New York

Library of Congress Cataloging-in-Publication Data
Gleick, James.
    What just happened : a chronicle from the information
frontier / James Gleick.
        p.   cm.
    ISBN 0-375-42177-7
    1. Telecommunication.   2. Cyberspace.   3. Computers
and civilization.   I. Title.
    TK5101.G58 2002   303.48'33—DC21   2001055443

www.pantheonbooks.com

Book design by Johanna S. Roebas

FOR CYNTHIA

[He] had that weird light in his eyes like he believed what he was saying the way some people believed in God or NASDAQ or the Internet-as-global-village.

— Dennis Lehane, *Mystic River* (2001)

CONTENTS

CONTENTS

WHAT JUST HAPPENED

INTRODUCTION

The last decade of the twentieth century came as a surprise. Of course we knew we had personal computers everywhere (suddenly). We knew we had a worldwide telephone network deploying vines and tendrils to every corner of the earth. We did not see that these facts—the computer and the network—were already a dyad. They were conjoining to make something greater. Many people were beginning to believe once again—as their great-grandparents had, when the century opened—that technology had the power to change life for the better. Not that fear and loathing altogether subsided. Indeed, as the calendar wound toward a number with three zeroes, we sometimes gave in to apocalyptic thoughts. Even in our calmest moments, our computers and our telephones inspired as much frustration as pleasure.

Sophisticated as the public was, it mocked George Bush (the elder) for his wonderment at the sight of a supermarket checkout scanner in 1992. He didn't care much about the Internet, either; nor, for that matter, did the brilliant young entrepreneurs at Microsoft.

In that year, the number of Internet nodes passed one million. The next year, the new president, Bill Clinton, acquired an e-mail address at whitehouse.gov. But few people knew these things. Mainstream newspapers would not print that obscure term, Internet, without adding an explanatory phrase like *global network of computers and their interconnections.* (I know, because from time to time I tried to use the word in the *New York Times.*) One year later, everything was different.

Something like this happened at the end of the previous century: a shift in the world's size and topology. When the new technologies of electricity and the telephone began to penetrate everyday life, people were amazed, confused, awestruck, and then—almost as rapidly—cured of their surprise. In the 1890s only a tiny fraction of the wealthiest and most futuristic Americans actually had electricity flowing to their homes, and their typical appliance inventory comprised a single incandescent light. They often had to unscrew the bulb if they wanted to plug in something else. (Wall outlets came along eventually.) But pioneers like General Electric and Western Electric and Westinghouse Electric already promoted a complex wired lifestyle, organized around the electric coffeepot (1891), the electric fan (1893), corn popper (1909), cigar lighter (1911), and alarm clock (1915).

The Bell telephone companies, meanwhile, were encouraging women to consider shopping by telephone: electronic commerce. At the turn of the century, even in California, which led the nation in telephone density, penetration had only reached about six households per hundred. But those new identifying coordinates, telephone numbers, were beginning to appear in advertisements and business cards. The word spread quickly; after all, communications formed the essence of this makeover of the landscape. People who were, one year, discomposed by the act of shouting "Hello" into a small mouthpiece became, in the next year, experts on the changing social patterns of their own lives. Loneliness, both urban and rural, was to be banished. In emergencies, help was only a phone call away. Time and space were annihilated, people said (and Einstein was just a teenager).

Of course, another technological surprise was rolling around the corner. The social historian Claude Fischer found this shocked headline in an Antioch, California, newspaper from 1906:

Many Deaths Due to "Devil Wagon"!

Dead and Unconscious Forms Left in Wake of Speeding Automobiles Operated by Reckless Chauffeurs.

So if we sometimes found ourselves, in the 1990s, alternating between shock and wonder, we weren't the first. We didn't quite see cyberspace until we got there; then again, our ancestors had glimpsed it, a hundred years earlier. "The news of the world is gathered from its four corners in less than a second's time," declared the president of the National Electric Light Association in 1896, and he was describing his present, not our future:

The softest whisper of the human voice is transmitted a thousand leagues . . . night is turned into day, darkness into light, the waste forces of nature are harnessed and wafted like spirits, unseen and instantaneously, over mountains and rivers, miles upon miles . . .

In reality, these transformations were far slower than they seemed. The diffusion of innovation took decades where now we think in terms of years. For both electricity and the telephone, roughly a half-century passed in the United States between commercial introduction and something resembling true ubiquity. Consumers of technology learn to be nimbler now—to skip lightly across the years, to pass with barely a tremor from vinyl to CD, from punch card to microdrive.

So I ask the reader, savvy as you are, to enter a time tunnel. Transport yourself back to the crepuscular past of the 1990s. We citizens of the twenty-first century have learned a few things, discarding our previous quaint mind-sets as snakes shed their skins. But we don't get away altogether clean.

When the decade began, I had been a science reporter, mostly. I was working to finish a biography of Richard Feynman. Succumbing to a classic form of writer's avoidance, I became more interested than was altogether healthy in one of my professional tools: my word processor. An adventure ensued. It is recounted in the opening essay, "Chasing Bugs in the Electronic Village." Apparently the electronic village is where I located myself, in 1992—in a "permanent floating conversation," a new culture in formation. Several of the themes that thread through this volume begin here: the problem of the software defect; the problem of the word *bug*, for that matter; the problem of Microsoft, and the problem of user angst. Now and again I expect the reader to be brought up short—as I am, rereading my words. How outlandish the past was! Was the dominant word processor really a thing called WordPerfect (made by a large, successful company called WordPerfect)? How strange to recall: "a relatively small number of personal computer users use Windows."

Someone spending time with scientists, at the universities and government laboratories, could hardly fail to notice their fondness for a new medium of communication. They exchanged letters electronically, without paper. They sent one another mail and the U.S. Postal Service was nowhere involved. This was no trivial business; it was shaking the fundamental structure of scientific communication. The grand refereed journals were not going to go out of business overnight, but, more and more, the real work of science—at least, the real *news* of science—transpired via electronic preprints. Occasionally a scientist would ask me, collegially, for my e-mail address. I was embarrassed. Where, on the Internet, was I?

I had wires coming into my house, transporting data, bidirectionally, kilobits at a time. It was analog data, and those wires had been there since approximately the time of Alexander Graham Bell. But I had a modem—a fast, 1,200-baud modem. I understood the meaning of bandwidth. I felt I should be able to connect to the Internet somehow. If I could just find it—this thing, this place, this entity . . .

A year later, I was researching an article called "The Telephone Transformed (Into Almost Everything)." We see now that this belonged to the time-honored journalistic genre, Something Is Happening Here But You Don't Know What It Is. I was trying to grasp a phenomenon that spawned cheap catchphrases like grass cuttings from a lawn mower: *information age . . . technological convergence . . . digital revolution*. The telephone seemed to be the device at the heart of things. It was, and it wasn't.

I began my reporting at the great research laboratories built over a hundred years of American telecommunications history: AT&T Bell Labs and Bellcore. Where else? Personal computer manufacturers were making their boxes; the network was not their department. Software companies had no evident interest in an emerging online world. There were commercial online services—mainly, CompuServe—but from inside their portals the Internet seemed as remote as the Magellanic Clouds.

The telephone companies themselves were notoriously staid—I quoted Nicholas Negroponte's assessment of them as "a bunch of sleepy janitors who haven't woken up"—but at least they had taken stock of the computing all around them, and some of the forward-looking thinkers at their laboratories had a vision of the future.

Actually, the head of research at Bell Labs, the eminent physicist Arno Penzias, tried to warn me off the Internet. In the big picture, he said, it was negligible—its total bandwidth barely a fraction of a percent of the vast telephone network whose command center I had just toured. In the spring of 1993, the Internet did seem an obscure piece of the story I was trying to tell. It was still obligatory to explain: "a disorganized, evolving entity built up from hundreds of thousands of interlinked university, corporate, and government networks." But, as I wrote, I couldn't help speaking again and again of the network, singular. The Network. An active, almost sentient creature. The Internet crept into the story through the back door and took it over.

Then I was a writer between books, and I thought I saw a way to create, for myself, at least, the kind of Internet access that the technical elite were beginning to take for granted. I started a company. It was called the Pipeline, and before it had two paid employees, it had a hundred customers. A month later it had a thousand, and Jack Rosenthal, the editor of the *New York Times Magazine*, asked me to tell him about it. I must have described my mixed wonderment and frustration—my sense of an anarchic and yet democratic frontier; my hope or belief that the future belonged to tiny mammals scurrying about at the feet of dinosaurs; my ambivalence about the gap between the promise and the reality—because, in assigning me to write an essay about my experience, he equipped me with this mantra: "I have seen the future, and it's still in the future."

That was the spring of 1994. I'm not going to tell the story of the Pipeline here. By the end of the year I was selling the company to get back to the writing business, but meanwhile I had learned some things that I had not understood, and might never have understood, as a reporter. As I wrote about technology and the future in the years that followed, I continued trying to fathom the emerging online world.

I got some things wrong—deliriously wrong:

- In "This Is Sex?" I claimed that the Internet was a poor medium for the delivery of pornography—invariably slow and disappointing—and that the government might as well stop worrying about it. I have to admit that online pornography thrived, instantly. It became an enterprise far more formidable and successful than I expected. Someone must be enjoying it. For entrepreneurs, it's even profitable. Well, for what it's worth, I still think that the Internet is a poor medium for the delivery of pornography and that the government may as well stop worrying about it, but apparently that's just me.
- In "The Information Future: Out of Control," I predicted the demise of the private online services, which were, chiefly, CompuServe and Prodigy, along with their smaller rival, America Online. "Those services may have a limited future anyway," I said. "Either they will open their gates to the Internet and become subsumed by it, or they will remain lakes isolated and apart from the ocean." Steve Case at America Online let me know that

he didn't appreciate being called a dinosaur and that he took the Internet very seriously. Weeks after this article appeared, he offered to buy the Pipeline. I never dreamed he would manage to buy everything from Netscape to Time Warner. Anyway, I was partly right: CompuServe and Prodigy flailed about for a while and are gone; and America Online did open its gates to the Internet, but I have to admit it hasn't quite been "subsumed."

By and large I tried to leave my crystal ball in deep storage. It's hard enough to understand the present; I didn't want to get into the business of prophecy, when the future was hurtling toward us with such disturbing velocity. Nevertheless, given the subject matter, a form of prediction became implicit in almost everything I wrote. None of us can avert our eyes from that shining horizon, can we?

So it's a relief, looking over these words two or three or ten years after they were written, to see that some things are still true. I was right about Microsoft. In 1995 it was generally believed that Microsoft was just a fast-growing software company, aggressive but law-abiding, with products that succeeded or failed according to their merit. *Successful* was a word more routinely associated with Microsoft than *powerful.* But the editors of the *New York Times* encouraged me to spend much of that year reporting an article for the Sunday magazine on the subject of power: how Microsoft acquired it, consolidated it, and wielded it ("Making Microsoft Safe for Capitalism"). We know what followed. The grand illuminating saga of *United States v. Microsoft* was still in the future.

Then, as 1999 began, it was generally believed that the turn of the millennium would bring catastrophe. Every major American news organization had assigned reporters to cover the Y2K "crisis" full time. Now it's a dim and perhaps comical memory, but it was no joke: people were stocking up on food, water, medicine, and ammunition. The federal government was demanding Y2K compliance reports from small businesses that didn't even have computers. Billions of dollars were drained from the economy. I thought this was nuts. Also, I had a guilty conscience: I worried that I might have contributed to the hysteria with my early, speculative essay, "Oh-Oh." So I attempted a Y2K antidote, "Millennium Madness."

The madness continued. With six weeks to go, the United States opened a $50 million crisis center. The state of Ohio actually moved its emergency government operations into an underground bunker eight miles north of Columbus. Hundreds of books were published: *How to Survive the Coming Crisis; Don't Be Scared, Be Prepared; Building Your Ark.* The *Y2K Personal Survival Guide* had instructions on "what to include in your home survival kit, how to purify your water, shelf lives of 45 common foods, phone numbers for your vital records for all 50 states, how to select a generator, and more!!" There were Y2K cookbooks.

Nothing is lost, in our online world (see "The Digital Attic"), so for amusement people can still look up these Y2K books at Amazon.com and read the reviews posted by earnest customers. The tone of the reviews changes sharply on January 1, 2000. My favorite reads, in its entirety:

I'm posting this on January third—nuff said.

We need to remember to check our collective pulse once in a while. Now that we communicate so fast and so well—*because* we communicate so fast and so well—we are more prone than ever to fits of mass hysteria. Information bounces from one channel to the next. Rumors feed on themselves. If it isn't Y2K, it might be El Niño or Monica Lewinsky—phenomena with some grounding in reality, but only some.

In a peculiar but very real way, Microsoft and Y2K are the same story. They're examples of what economists have started calling "network effects." A law of increasing returns governs them, not the law of diminishing returns. They benefit from positive feedback. The more people [use Microsoft software | worry about Y2K], the more pressure there is on others to [use Microsoft software | worry about Y2K]. And this is independent of the actual [quality of Microsoft software | danger from Y2K].

By the way, I was right about patents, too (another heterodox polemic: "Patently Absurd"). Time will tell. The networked world isn't just a little bigger, a little faster, and a little more complex. It's qualitatively different. We need new psychology, we need new economics, and in specific cases we need new public policy.

I see now that this book is a credo. I may not believe in God or the NASDAQ, but I do believe in the network as global village and as global brain. I believe in privacy—and not necessarily in anonymity ("Big Brother Is Us"; but "Multiple Personality Disorder"). Notwithstanding the dangers of tidal hysteria and cacophonous discord, I believe that communication is inherently good. As for commerce: let it be an afterthought. From the start of this decade to its finish, I cared more about how Amazon intuits my likes and dislikes than about whether it would ever make money. I seem to have cared more about a single hot tub that went online in Ypsilanti, Michigan, than about the entire meteoric arc of

Netscape, Inc. The Internet is not a shopping mall, and, for that matter, the World Wide Web is not the Internet. The reader will see that I never put faith in the dot-com boom; nor am I troubled by the dot-com crash. I believe that access to the network must be universal, for the sake of social health and political sanity. Like water, like electricity, like the mail, the tools and standards of the digital world must be open and public.

I still believe in the 1937 prophecy of H. G. Wells, which I cited in "Here Comes the Spider"—and never mind that he was talking about a different, trivial and now obsolete technology: "It foreshadows a real intellectual unification of our race. The whole human memory can be, and probably in a short time will be, made accessible to every individual. And what is also of very great importance in this uncertain world where destruction becomes continually more frequent and unpredictable, is this, that . . . it need not be concentrated in any one single place. It need not be vulnerable as a human head or a human heart is vulnerable."

He had seen the future, and it was still in the future. As for us—it's a brand-new century. And we're still at the stage of unscrewing our solitary light bulb whenever we want to plug in the fan.

—J. G., February 2002

CHASING BUGS IN THE ELECTRONIC VILLAGE

August 1992

I couldn't wait to buy Microsoft Word for Windows—rumored to be the new Cuisinart, Mack truck, and Swiss Army knife of word processing software, full-featured, powerful and, for a writer, the ultimate time-saving device. I was writing a long book, and I wanted the best. One day in January 1990, I finally got to tear open a software box bigger than some computers, and out it came. The world's preeminent software manufacturer had spent roughly as long developing this word processor as the Manhattan Project had spent cooking up the atomic bomb, but secrecy had not been quite as airtight. For more than a year, Microsoft had been leaking juicy tidbits to its waiting army of trade journalists, computer consultants, and corporate purchasers. Word for Windows (aka Winword or WfW) would be Wysiwyg (the standard acronym for What You See Is What You Get)—that is, it would display page layouts and typefaces with high fidelity to the final printed product. It would let users work with nine documents on the screen at once. It would have a macro language—a way to spend hours writing mini-programs to streamline all those little chores that can suck up milliseconds of a writer's time.

And it would also have—in some corner of my mind, I must already have known this—bugs.

Computer software is the brightest of bright spots on the American economic landscape, a consumer product evolving in a floodtide of innovation and ingenuity, an industry that has barely noticed the recession or seen any challenge from overseas. Bugs are its special curse. They are an ancient devil—the product defect—in a peculiarly exasperating modern dress. As software grows more complex and we come to rely on it more, the industry is discovering that bugs are more pervasive and more expensive than ever before. Word for Windows had big bugs and little bugs. A little bug might mean that a user would sometimes find the em dashes (—) and en dashes (–) switched. Even a little bug could send users running for their blood-pressure medication. "This bug is severe," one user railed at Microsoft when he discovered what was happening to his ems and ens. "It renders the whole setup useless for any serious work."

A bigger bug would cause unwanted typefaces to appear willy-nilly in one's documents. An even bigger bug would cause a poignant message to appear on screen

Unrecoverable Application Error
Terminating Current Application

after which the computer would crash, die, freeze, lock up or hang (the slang was evolving too fast even for Word's electronic thesaurus to keep up). The U.A.E., as it became known, would send one's current work into the digital oblivion so familiar to computer users.

16

• • •

Searching for help, I stumbled into an odd corner of the electronic village, a "forum" on the CompuServe Information Service devoted entirely to a permanent floating conversation about the ins and outs of Word. CompuServe is a vast electronic network that subscribers can dial into via modem to use a wide array of services from game-playing to stock quotations. Twenty-four hours a day, users from a dozen time zones dial in, read their mail and everyone else's, and post replies. Microsoft's Word for Windows forum provided instant access not only to the experiences of other users but also to Microsoft's development team—because an assortment of programmers also joined in, including the program manager responsible for Word's development.

CompuServe fosters a strange form of communication, more casual than letters, more formal than telephone conversation, and extremely public. An electronic culture has developed with its own evolving rules of decorum, its own politesse and its own methods of conveying rage, not to mention its indispensable acronyms: **IMHO** (in my humble opinion), **FWIW** (for what it's worth), **PMJI** (pardon my jumping in), **ROFL** (rolling on the floor laughing), **IANAL** (I am not a lawyer). The characters **;)** are supposed to mean "just kidding"—on the theory that they suggest, sideways, a face that is smiling and winking.

The medium creates a modern kind of town meeting—perhaps the only real town meeting left—where opinions from the humble to the slanderous fly freely. When war breaks out in the Persian Gulf, say, CompuServe instantly opens a new forum. But never mind war—these days the waters are roiled by debate over the rivalry between Microsoft and IBM, with partisans as fierce as any religious zealot. "I have never seen such a vicious crowd," one CompuServe participant said recently, and another replied: "Hum, try the Issues forum Paranormal section where the Bay Area Skeptics and the psi-clones are mugging each other." I began spending hours each week watching Microsoft's product forum as users discussed Word's fine points, traded tips like fussy chefs exchanging recipes, and harangued the company's ever-diplomatic support crew about the essential mystery: Why couldn't the software be cleansed of all these bugs? Every so often there was progress—of sorts.

"Your vociferous complaints, while a bit overbearing, have borne fruit," the program manager, Chase Franklin, told me. "I just met with the testing engineer assigned to the problem and he has determined two things: The model 80 he was using did not hang when displaying the symbol fonts, it just took about three minutes to update the screen to a correct display, sans the symbols." Translation: My latest bug was not actually crashing the computer, as I had thought. It merely caused the computer to pause for three minutes after each character I typed.

Without quite meaning to, I had become a member of a nettlesome group of users that someone dubbed the Winword Gadfly Team. The other Gadflies were a Hollywood screenwriter teaching at Columbia University, a Toronto teacher, and a Caltech mathematical physicist. We loved our word processor—it let us fiddle and tinker until we had completely automated tasks as simple as counting the words in a chapter or as complicated as formatting a screenplay. And we hated it—because the more we used it, the more bugs we discovered. As we never tired of pointing out.

"Don't you guys ever work?" an exasperated Microsoft developer asked a few months later. "I swear you live here."

· · ·

Software bugs defy the industry's best efforts at quality control. Manufacturers may spend far more time and resources on testing and repairing their software than on the original design and coding. "Debugging" is not just an integral part of the development process; it is sometimes the dominant part. Programmers are trying to combat the increasing complexity of their creations with new techniques—modular design, for example, that might contain damage as flood compartments do in a ship—but so far these have made little difference.

The problem is that software is different from other merchandise. Computer programs are the most intricate, delicately balanced and finely interwoven of all the products of human industry to date. They are machines with far more moving parts than any engine: the parts don't wear out, but they interact and rub up against one another in ways the programmers themselves cannot predict. When a program doubles in size, the potential for unexpected bugs more than doubles—far more, just as the number of potential love affairs more than doubles when the population of your office rises from ten to twenty.

So developers turn to user testing. Beta tests, as these adventures are known, have grown to enormous scales. Microsoft and IBM have each recently concluded beta tests in which tens of thousands of users participated. The Word for Windows beta test had lasted many months—but not long enough, it seemed. We customers began to feel like unwilling beta testers. And perhaps not Microsoft's favorite beta testers at that—certainly I wasn't doing my part to maintain the upbeat spirit of the company's CompuServe forum.

"When I said, 'No no no no a thousand times no' just now, you might have interpreted that to be sort of, well, negative," I conceded to one developer. A year had passed since the original January 1990 release, and strange typefaces kept showing up here and there. Some users were tempted to give up Winword altogether—but changing word processors is almost as traumatic as changing religions, and Winword's competitors had their bugs, too. Meanwhile, to temperamental writers, hoping that their word processors would become as second nature as an old typewriter, every encounter with a bug was a slap in the face. Guy Gallo, the Columbia screenwriter Gadfly, encountering the typeface problem yet again, hit his Caps Lock key: **"THIS IS THE SAME BUG WE HAVE BEEN SCREAMING ABOUT SINCE WFW'S RELEASE."**

That was the problem. Winword 1.0 had been updated with Winword 1.01, Winword 1.1, and then Winword 1.1a. (Microsoft was doling out version numbers sparingly.) A few bugs had been eliminated but more had been discovered, and the general impression was of sliding backward downhill. Why were we still talking about bugs that had been reported and confirmed a year before?

"You need to read your license agreement," one of the developers declared on CompuServe after I had needled him for a while. "We don't have an obligation to issue another release of the product, James, and it's warranted on an 'as is' basis. We don't even have a legal obligation to worry about your data loss." This was true. Some bug-plagued software users have tried to sue manufacturers for damages, but the courts virtually never sustain such claims.

The Gadflies discovered that Word's formatting instructions did not function as expected for footnote numbers. Worse, Winword's "templates"—documents that stored styles, customized command menus and other user information—seemed to take an increasingly long time to save. At first they had saved in seconds, like normal documents. As the templates grew larger, however, they seemed to cross an invisible threshold, and now users found themselves waiting two minutes or more—long enough to panic and reach for the on–off switch. Worst of all perhaps, typefaces—fonts—behaved in a variety of weird ways that someone gave the name "phont phunnies."

The developers said they were trying. It was a hard task: users would report problems that Microsoft's testers could not reproduce on their machines. Different computers, different amounts of memory, different documents, different combinations of software made it impossible sometimes to track down bugs, although the Winword program manager said he now had twenty test engineers tracking our reports.

But Microsoft's marketing strategists had more pressing problems. Winword was by far the dominant word processing program designed to be used with Microsoft's Windows operating environment, but a relatively small number of personal computer users use Windows. Overall, the market leader by a large margin was WordPerfect, which was known to be beta testing its long-awaited entry into the Windows market. Microsoft officials were worried. WordPerfect commands enormous loyalty, in part because—unlike Microsoft—the company makes a practice of releasing frequent free upgrades to repair even minor bugs, and in part because it maintains, at enormous expense, a toll-free telephone support line—an investment Microsoft, which says it fields 14,000 calls a day, has been unwilling to make.

•   •   •

In June 1991 the Gadflies arrived en masse at Microsoft headquarters, in Redmond, Washington, a suburb of Seattle. The corporate giant seemed to have disguised itself as a small college. Its bucolic 256-acre campus mingled elegant plantings with outdoor basketball courts. Its executives and developers turned out to be mostly in their twenties, wearing jeans and T-shirts. It was as if gray hair, neckties, and the words *Mr.* and *Ms.* had been banished from the earth.

Plucked from our faceless electronic existences, we had been flown in for a preview of Word 2.0—to serve as a sort of cranky focus group, we supposed. Curmudgeons, meet your new word processor. Finally, here was the major upgrade: Microsoft hoped it would keep WordPerfect from achieving a new dominance in the Windows market; we hoped it would rescue us from the bugs.

We loved it. We hated it. Winword had amazing new features. Not just the beautifully designed "mail-merge" feature—lone writers don't have much occasion to send customized mass mailings linked to thousand-entry databases. Not just the "grammar checker"—I already know my sentences are too long, thank you very much. There was an automatic envelope printer— Microsoft's research had revealed that some people still kept typewriters on hand for that purpose alone. There was a built-in drawing program, a built-in business chart program, a built-in equation editor. If I wanted to number a group of paragraphs, insert a table, zoom in on my text, or create side-by-side columns, I had merely to click my mouse on a pictorial icon at the top of my screen. No wonder my book was already a half-year late.

Yet there were signs that all was not well. A feature meant to compare two versions of a document still did not actually function; it was a shell, as one of the developers admitted privately, with little more purpose than to persuade the trade press to add one more "Yes" to the feature-comparison charts that always accompany word processor roundups. ("There were so many higher priority items that users requested, we couldn't squeeze it in," Microsoft says now, adding that the feature would "probably" work better in a later version.) The strange font problems seemed to remain—but perhaps it was too soon to tell. Another beta test was getting under way.

For the Winword team, the next months brought intense pressure. WordPerfect for Windows was rumored to be imminent. WordPerfect had run a long, large beta program. Microsoft's programmers had been able to monitor its progress and thought they could match it feature for feature. Still the original, non-Windows version had 9 million or 10 million users who would naturally be inclined to stay with a brand they knew. And meanwhile, another competitor, Lotus Development Corporation's Ami Pro, came out with a startlingly improved new version.

Winword 2.0's release slipped from September into October. The Gadflies, their ranks swelled with new volunteers, were shocked—shocked!—to discover that the developers were deliberately leaving bugs unrepaired: they were declared to be "features," or "by design." It was triage. In a candid moment on CompuServe, Microsoft programmers admitted the existence of an internal list called, evocatively, "Won't Fix."

Had twenty months of angst-ridden electronic messages been for naught? Guy Gallo begged for some assurance that "Won't Fix is more like Bug Purgatory than Bug Hell." At a Windows conference in August, one of the newer Gadflies, Ellen

Nagler, a California consultant, handed out buttons: **Won't Fix** and **It's Not a Bug—It's a Feature**. The Winword developers asked for a set of the buttons. Emotion ran high.

"It IS only SOFTWARE," one of the developers typed in exasperation.

Finally, in November, the program reached stores, accompanied by an enormous wave of promotion that included, for the first time, commercials on network television. Reviewers raved. They loved the ability to drag and drop chunks of text with a mouse; they admired the new mail-merge and envelope-addressing features; and, sure enough, they fell into Microsoft's trap and gave Winword a "Yes" in the document-comparison check box. They barely mentioned the bugs. How could a reviewer on deadline be sure that any particular problem wasn't an example of another chronic and costly industry problem: the one known as User Error?

The CompuServe message traffic was not so polite. Users reported old bugs and new ones. It seemed that you could crash the program by using the spelling checker on certain parts of documents. The feature that let you search for a certain word or phrase seemed to break under peculiar circumstances. Users who mixed pictures with their text found that the pictures tended to disappear when the document was printed out. Microsoft's CompuServe support staff scrambled to field the complaints. "EIGHTEEN MONTHS LATER, this is approaching ridiculous," one longtime user exclaimed. "Explain to me again how, with all that feedback, and with all the input from the Gadflies before and during the beta, these major hassles never got fixed? What's it gonna take?"

Said another customer: "I'm dead in the water and so are all of my users. Quite frankly we are up in arms about this. Why doesn't MS debug the program before releasing the product to the public. . . . Help me, my ship is sinking!!!!"

In a simpler era, two or three years ago, software vendors could bring a programmer to the telephone, make a quick repair to their code, and ship out a replacement diskette within days. But now every change in a program must be tested exhaustively to uncover possible side effects. At best the company manages to offer a maintenance upgrade a few times a year.

Microsoft, accustomed to gentle treatment by the trade press, refuses to make its top executives available to discuss its policies on software defects. Chris Peters, the general manager for word processing applications, says carefully when asked about the persistence of bugs, "The goal is not to have any." He inadvertently fueled the battle raging on CompuServe by telling the press that only a tiny handful of advanced users would ever encounter flaws in the product; meanwhile, his programmers were working to eradicate some of them in Winword 2.0a, finally released this March. Yet even now, as I type away in a standard font called Times New Roman, if I copy some ordinary Times New Roman text from another document, I will see . . . wait . . . horrors! . . . an unwanted font that Winword labels Futura though it looks suspiciously like Courier.

Gradually the Gadflies have tried to get back to work and finally take advantage of our state-of-the-art writing tool, our ultimate time-saving device. But new users continue to turn up, complaining about weird font changes, little realizing that they are seeing a bug with two years of history. Microsoft's support representatives tell them politely that their suggestions will be considered.

Gadfly echoes still reverberate through the electronic village. The other day I noticed one user telling another: "You know who you should contact is James Gleick. He really must have a love–hate thing with WfW, because the level of frustration he seems to experience would kill a normal man."

THE TELEPHONE TRANSFORMED
INTO ALMOST EVERYTHING

May 1993

When the telephone rings in Arno Penzias's office at AT&T Bell Laboratories, nothing is simple: first comes an electronic beep from one of the button-studded, silicon-filled instruments on his desk; then the cordless handset on the conference table trills along behind. Penzias, the Nobel laureate who directs research at this most famous of American corporate laboratories, grouses: Why didn't the caller use electronic mail instead? Voice is so retrograde, so intrusive. Penzias gets up and switches phones to demonstrate one of his latest toys, something the Bell Labs engineers call "Bump in a Cord." It is an eavesdropper-foiling encryption device the size of a paperback book, and it works so well, he says, that the government will not allow it to be carried out of the country ("Princess Di would love one"). But you do need another Bump at the other end; for a proper demonstration, Penzias must look up a colleague's phone number. So now, for just a moment, we put the information age on hold while he opens a drawer and rummages through a sheaf of what looks suspiciously like paper.

Paper?!

"Oh," he says. "You mean why no voice dialing? Saying 'Get Fred' or 'Get Sam' or something? I don't have a voice dialer." He will soon, surely. Any day now, he will be able to pick up a phone anywhere and say "Call home" to a network that will recognize his voice and look up his home number (verifying his credit by means of his voiceprint). If in the meantime Penzias, like most people who own phones with programmable buttons, does not keep them programmed, it just goes to show that nothing in the telephone universe is simple.

Well into its second century, the telephone has begun a transformation more profound than any in its history—dragging along with it much of our other technological baggage, including the computer, the fax machine, the pager, the clock, the compass, the stock ticker, and the television. In the past year it became possible to rent cellular phones with your skis and poles at Vail, Colorado, and from vending machines in California. The last three area codes assigned to North America were used up—assigned for the "relief," as the phone companies say, of over-numbered regions of Philadelphia, Michigan, and North Carolina. More transatlantic circuits have been added in the past three years than in all previous history. In Portland and Seattle you may arrange to receive phone messages, along with stock prices and basketball scores, on a wristwatch. You need no longer visit the Western Wall in Jerusalem to stuff your note into a crack; the Western Wall now accepts messages by fax. Perhaps that was to be expected. But who would have thought that the downfall of the English monarchy, echoing through the ether, would have sounded like this:

CHARLES: Oh, God. I'll just live inside your trousers or something. It would be much easier.

CAMILLA: What, are you going to turn into a pair of knickers? . . . Oh, darling, don't be so silly. I'd suffer anything for you. That's love. It's the strength of love. Night night.

Is the technology about to liberate us or to overwhelm us? Like Charles and Camilla, we have tended to take our telephones for granted. In contrast to other great modern life-altering technologies—the automobile, the television—the telephone, until now, has not inspired much in the way of hobbyists, collectors, popular magazines. When it rings, we answer. Meanwhile, we don't watch it, polish it, or even read about it.

But here comes a glossy new magazine called *Wired,* announcing ecstatically, "The Digital Revolution is whipping through our lives like a Bengali typhoon—while the mainstream media [are] still groping for the snooze button." *Wired* stands ready to discuss the meaning or context of "SOCIAL CHANGES SO PROFOUND their only parallel is probably the discovery of fire." Industry executives and federal regulators use language only a few shades less apocalyptic. "All this accumulated technology and accumulated vision," says Stewart Personick of Bell Communications Research (Bellcore), "is like a volcano waiting to explode."

They see a converging of technologies and industries that have been independent until now: most notably the computer, large and small, and a half-dozen cousins of the basic telephone.

Just as the fax machine—which has existed as commercial technology since the Civil War—finally came into its own all at once in the late 1980s, so a whole host of other telecommunications devices and networks suddenly seem to be achieving the necessary critical mass of users. When AT&T's chairman, Robert E. Allen, gives a speech, all his graphs are the same—a trend line soaring toward the (presumably blue) sky. Only the legend changes: Networked Computing, Wireless, Messaging, Visual Communications, Voice and Audio Processing. . . . (He might as well add Jargon and Acronyms—the entries in *Newton's Telecom Dictionary* have ballooned through four editions in five years, the last growing by 35 percent.) Indeed:

- The United States population of cellular phones—handheld or in cards—grew last year by 46 percent, after growing 43 percent the year before. Cellular companies now report 11 million subscribers.
- Pagers—displaying the caller's telephone number or, more and more often, words and sentences—are spreading just as quickly. There are more than 14 million out there, with personal use rising more rapidly than business use. The paging companies hope that parent–babysitter and teenager–teenager communication will never be the same.
- Online information services, having begun with hardcore computer hobbyists, also report huge growth rates. About 20 million people use the many networks linked by the noncommercial Internet, exchanging electronic mail, joining thousands of grass-roots mailing lists, searching databases and catalogues or playing long-distance interactive games.

These technologies are merging in little science-fictiony boxes that are not quite computers and not quite telephones. Apple Computer has its Newton, and AT&T and EO have their Hobbit personal communicator; Motorola, Sharp, IBM, Tandy, Casio, and many others have bruited their forthcoming "personal digital assistants" or "pocket pals." It is a highly vaporous category of product, the glossy photographs and nonworking prototypes having long preceded real merchandise. Still, they seem to be on their way. And they represent a serious blurring of the traditional industry lines: the days when computer companies processed data and telephone companies placed calls are gone forever.

These little boxes will be connected, one way or another, to that vast, entangled, amorphous creature known to those in the business simply as the network. The Clinton administration has made the creation of a national "information superhighway" one of its watchwords, but the superhighway is already there and growing at a phenomenal pace, with new fiber-optic and wireless pathways announced weekly by established telephone companies, cellular entrepreneurs, and empire-building cable television companies.

To say that questions of privacy are involved would be to miss the point—our sense of what can be private and what must be public is being overhauled under our noses. The administration has set off a furor in telecommunications circles with a new proposal to build into the electronic infrastructure a form of encryption designed to make government wiretapping not only possible but also efficient. The network already has a frightening kind of intelligence built in, and soon it may know at any instant where you are, who you are willing to take a call from, how your credit stands, what kind of restaurant you like, how you can get there, and whether you'll need to bring a raincoat. Smart terminals in a smart network—that is the central image.

"All these devices that the major manufacturers have announced," says Steve Brendle, a Motorola executive, "—you almost have to believe that all these major companies can't be wrong, and a couple of years from now we're all going to be carrying around this thing, whatever it is. We'll take notes on it, and we'll also receive messages from people, the special of the day from the grocery store, the daily calendar of events for a politician . . ." Brendle's own project is the newly announced Embarc—an acronym (inevitably, in the telecommunications business) for Electronic Mail Broadcast to a Roaming Computer. With the help of a satellite radio network and a tiny receiver, subscribers can have an ongoing stream of information purr into their computers or their pocket pals—phone numbers of callers, e-mail, physicians' lab results, *USA Today* news bites, daily company sales figures.

"People are frightened by the thought of getting too much information, which just shows we're not in the information age yet," says Penzias at Bell Labs. "Are you frightened by the thought of getting too much money? Too much happiness?"

We have reached the brink. "On one side, you still have the Age of Paperwork," Penzias says. "On the other side you have something I like to think of as information transparency.

"And when will you know you're in the Age of Information Transparency? I will tell you. If somebody says I can get you ten times as much information as you have now, if that makes you feel good, you're in the Age of Information Transparency. When people relish the thought that the information flows to them can become bigger, then you're in the Age of Information Transparency. And until then we're in the Age of Paperwork."

Decades of talk about the information revolution have not

spared us traffic jams, paper shuffling, or endlessly ringing telephones. How we communicate will depend less on the visions on
display in the laboratories—startling though these are—than on
the human abilities and desires caught up in the blending of past
and future. Will we *want* to carry those boxes and wear those
watches; will we triumph over the Switch Hook Flash and the
"voice-mail jails"; will we, intentionally or not, drift into new
forms of community?

We are noticeably self-conscious about the unshackling of our
phones. As they move from bedside tables and kitchen walls into
our pockets, we still react with degrees of curiosity and annoyance to the sight of people using cellular phones in public places.
A plane delay at La Guardia airport: people nudge and stare as a
man in a suit shouts into a glossy black wallet-sized phone, "No, I
can't, my battery's about to run out." A bank of pay phones is
available fifty feet away. A letter in *PC Magazine* castigates those
"unfortunate boors who will continue to flaunt their inflated
self-importance with flashy doodads in the same way they wear
their ostentatious jewelry." In one recent *New Yorker* cartoon a
drowning man says into his cellular phone, "Help!"; in another a
man kneels to propose to a woman who says into her cellular
phone, "Dave, could you hold on a sec while I take care of some
personal business?"

We will use telephones as lifelines even more than in the past;
we will have to grapple with new questions of etiquette and social
propriety. Will the public use of cellular phones remain a sign of
ostentation and antisocial rudeness? Or will it come to seem as
normal as the retreat to a pay phone or the use of a Walkman on
the bus?

Either way, the telephone, seeming recluse among information appliances, has stepped forward. Marshall McLuhan noticed a generation ago that the telephone was never really an instrument of the background, like the radio or the television. It has never addressed us passively; rather it has engaged us forcibly from the start: "an irresistible intruder in time or place," he wrote—"with the telephone, there occurs the extension of ear and voice that is a kind of extrasensory perception." All the more so now that it can be as portable as a wristwatch. McLuhan cited the eerie, classic instance of what he called its "cooling participational character," the case of the 1949 mass murderer Howard Unruh. Unruh had shot sixteen people with a war-souvenir Luger on the streets of East Camden, New Jersey, and then holed up in his house. Police trained machine guns on the building, a tear-gas grenade shattered a window, and the telephone rang. Unruh, bleeding from a wound in his thigh, picked up the receiver and said hello. It was an enterprising local newspaper editor. The *New York Times* the next day reported the following conversation:

This Howard?

Yes, this is Howard. What's the last name of the party you want?

Why are you killing people?

I don't know. I can't answer that yet. I'll have to talk to you later. I'm too busy now.

Telephones cannot be ignored. New models of cellular phones and papers can be set to vibrate soundlessly against our skin. NEC and Motorola have begun shipping phones that weigh less than half a pound. The boom in these technologies has sustained a fast-growing industry that seemed to ignore the recession. Although the average monthly cellular bill dropped last year to just below $70, total revenues rose almost 40 percent from the year before, to $7.8 billion. The number of cell sites, the basic call-broadcasting centers, has grown from 300 nine years ago to more than 10,000 at the end of last year. The cellular industries are rising rapidly overseas as well, particularly in places like Eastern Europe, India, and China, trying to modernize their phone systems more rapidly than a cable infrastructure can be laid or repaired. All over Hungary, where telephones in homes have been rare, the fast-spreading local cellular network has been broadcasting the slogan, "Put down your bicycle—and use your Westel telephone." When a cable break cut off normal service to the Radisson-Slavyanskaya Hotel in Moscow last winter, guests received cellular phones at heavy discounts.

Like so many of the agents pushing telecommunications into the future, the cellular industry suffers confusion over competing standards and shifting industry alliances. The central problem is simple: How to turn a hodgepodge of local systems into a seamless worldwide network. Users now need to think about where they are and where they are calling; they need to memorize access codes and programming instructions. It is proving surprisingly difficult to consummate the natural marriage of cellular phones with cordless phones—cordless being more workable inside buildings but limited to a very small range. The first effective cellular-cordless hybrid has just reached the market—a joint venture of Southwestern Bell and Panasonic—and it requires the phone to alert a central computer whenever its owner leaves or enters the building.

For now, with the best cellular phones, the ability to place calls, and to hear them once they are placed, fades in and out as people move about. "Microcells"—low-power miniature cell sites—have begun to broadcast inside buildings and tunnels. Regulators have approved new digital radio networks to link these sites without wires.

Computer companies, local phone companies, and cellular companies have formed a bewildering array of new alliances and joint ventures: McCaw has joined Southwestern Bell; Ameritech, Bell Atlantic, and Nynex have joined GTE in an effort to offer mobility—"roaming"—throughout 130 local systems. AT&T has bought a giant chunk of McCaw. BellSouth, Intel, Ericsson, and RAM Mobile Data are working on a communications network. McCaw has bought pieces of LIN Broadcasting, American Mobile Satellite Corporation, and Claircom Inc.—itself a Hughes Network Systems joint venture offering service to aircraft. McCaw and IBM, IBM and Motorola, Motorola and Bell Canada are all forming teams meant to extend their networks. Sorting out collaborators and competitors has become impossible, and federal antitrust officials are watching mainly with a detached fascination.

Overall, the mesh of regulation by the Federal Communications Commission, other federal agencies, and local public service commissions has become an intricately overgrown jungle. "The regulatory environment is rather nonintuitive," Personick at Bellcore says mildly. "If you explain it to someone, they figure you couldn't possibly have meant what you just said." The psychology of both the regulators and the regulated is dominated by one overwhelming event, the 1982 breakup of AT&T, now seen even by the company as a liberating historical moment. The Modification of Final Judgment is spoken of with the respect English historians reserve for the Magna Carta.

As AT&T transforms itself, as it seems, from stagnant monopolist to creative entrepreneur, the heirs to the Enemy of Progress mantle are the local phone companies—"a bunch of sleepy janitors who haven't woken up" is a typical comment by Nicholas Negroponte, head of the MIT Media Lab. They, in turn, argue that they are handcuffed by the terms of the breakup and by local regulators, whose main priority is to ensure cheap continued basic service for everyone who wants it. It is typical of the crisscrossing of regulations that phone companies are forbidden (by the 1984 Cable Act) from competing with cable television companies in their local areas—but not elsewhere. So Southwestern Bell, for example, has just bought two cable television companies in the Washington suburbs, opening what seems likely to be an era of violent leapfrogging. Nothing prevents cable television companies from offering telephone service on their cables, and they seem sure to begin doing so. Ready or not, local phone companies are about to face the first intense competition in their history. A wide range of alternative providers will shortly be offering to let you, as the expression goes, "buy dial tone."

Meanwhile, a drive-through cellular telephone store has opened in West Los Angeles; customers sip cappuccino as they wait for their new phones and number. Airlines now follow their seatbelt instructions with a routine warning against cellular-phone use—and phone enthusiasts debate whether this reflects legitimate fears of electronic interference or a desire to protect Airfone revenues. Just as parents have begun to outfit their children with pagers (because who can bear to be out of touch when it is possible to stay *in* touch?), schools have begun to ban them—

because pagers have become so notorious a status symbol for drug dealers.

"The landscape is changing unbelievably fast," says David Rose of Ex Machina, a New York software company. One of his products will automatically deliver anything that happens in your computer to a pager anywhere in the country: an alarm triggered by a spreadsheet, a new piece of electronic mail. "Motorola, for example, is a fairly staid company, but now they've got pagers that play wild space sounds! Pagers that are a fluorescent green!"

Appurtenances like answering machines and e-mail boxes, at first resisted as antisocial, are not only coming to seem essential; they are subtly changing our language and our manners. Ruby Gold, a researcher at the University of California Center on Deafness, has analyzed samples of "answering machine talk" and found what she sees as a new style of discourse. We have to learn the odd skill of speaking freely to someone who won't hear us until later. And when we do learn, some of us manage to crate monologues with extraordinary whimsy or passion. "It has evolved into this whole answering machine culture," she says. "It has really changed people's ability to do this funny thing, of speaking with an absent interlocutor." Etiquette and technology get tangled. Movie fans have been arguing via electronic mail about whether a phone operator should have been able to help the hero Jack Ryan in the movie *Patriot Games* when he tried urgently to break into his wife's cellular conversation so he could warn her of danger approaching her car. As the telephone lurches toward complete portability and perfect ubiquity, we are barely beginning to see the consequences for business, crime, or romance.

It used to be that just having a phone was enough. "What's

up?" said one cowboy villain to another in a 1950s tête-à-tête on
The Roy Rogers Show.

> **Plenty. According to Frank, Rogers has**
> **been tryin' to find a telephone. Says he's**
> **been lookin' around for one for the last**
> **couple o' days.**
>
> **If Rogers gets to a phone, we're in trou-**
> **ble!**

Now we get into more complicated kinds of trouble.

"GRIFFIN'S RANGE ROVER comes to a stop in front of
the house," wrote Michael Tolkin in his screenplay for *The Player*
last year, beginning a sinister and lovestruck scene possible
only in the cellular era—a blending of phone romance with
voyeurism.

> **GRIFFIN checks the phone number and**
> **dials it from his mobile phone . . . The**
> **phone rings for quite a while before JUNE**
> **seems to hear it. GRIFFIN gets out of his**
> **car and approaches the house with his**
> **phone in hand. He moves right up to the**
> **front window and watches her. He can hear**
> **the phone ringing both inside the house**
> **and on his phone . . .**
>
> **JUNE: Yes? Hello?**

•   •   •

"I carry around one of these little boxes," says a Motorola executive, Al Zabarsky, "and every day, besides my personalized mail, I get clippings from services that clip according to your temperament, from companies that specialize in original source material—whatever you guys in the business call it."

We used to call it *news,* Al.

With changing telephones come altered relationships to information of all sorts. Someone who logs onto the Internet has instant access to something else called news—news in its rawest and most ancient sense, messages broadcast or sent one-to-one, in thousands of groups. A group may be `alt.fan.dan-quayle` (where a long-running series of messages still spins off the topic "Mr. Potato Head") or `rec.music.indian-classical` or `alt.books.anne-rice` or `all.binaries.pictures.tasteless`—not to be confused with `alt.binaries.pictures.erotica`. The Internet—a disorganized, evolving entity built up from hundreds of thousands of interlinked university, corporate, and government networks—remains far from user-friendly, compared with commercial services like CompuServe and Prodigy, but it seems to be reaching a watershed. Non-technically-minded individuals without university connections can now attach themselves to the Internet through commercial services that are springing up. There is an astonishing breadth of knowledge here, and an astonishing ability of people to find sympathetic sharers of the most esoteric interests. And no matter how long we remain online reading such news, on exit we get a reminder that we are drinking from a deep sea:

> `There are still 1,038,071 unread articles`
> `in 1,388 groups.`

The business of providing information—gathering, reporting, editing, organizing, and disseminating news—has barely begun to respond. One service traditionally provided by institutions like the *New York Times* has been the sifting of information: editors use the limited space and varied typography of a newspaper to express judgments about what news is important. Their judgments are broadcast in more or less identical form to a million readers. Big newspapers and television news shows are culturally unifying, but, with a few exceptions, these organizations have been slow to adapt to the changing information marketplace. And cultural fragmentation seems to lie at the heart of the information services springing up across the network.

At Bellcore, the wing of Bell Labs that spun off with the Baby Bells after the AT&T breakup ten years ago, two of the many prototype systems for replacing human judgment about what information we do and do not want to receive are called Simple Electronic Filtering Tool (SIFT) and Electronic Receptionist (no acronym to date). SIFT is designed to sort incoming mail and news according to personal preference, forwarding it to one's pocket pal, or e-mail, or voice mail (after automated conversion to speech), or fax machine, or wastebasket. There are reasons to worry about such systems. One is that they require a huge amount of machine intelligence; a computer trying to identify your love letters is liable to misclassify **I'd love to see your sales figures** and **If I can't see you I'll kill myself.** Another is that they seem to drain the element of chance from our everyday wanderings through the world of information—the news we stumble on when turning pages.

There is surely a kind of obsessiveness at work in the accumulation of gadgets and the information service subscriptions that let us have news bits instantly around the clock. **These days you can't afford to wait for business news,** goes a pitch for the Prodigy online network, though it might as well be for any of several dozen services. Who can afford to wait? "There's a tremendous drive toward increasingly fast gratification," says James Katz, a sociologist at Bellcore. "People feel more time-stressed." For most of the new technologies faster means smaller: headlines and sound bites. Al Zabarsky still enjoys the nostalgic, tactile feeling of reading his Sunday paper by the swimming pool, but the kind of news that you *can't afford to wait for* tends to be quicker to digest as well as deliver.

Still, everyone seems to be going online.

Physicians are setting up programs to transmit not just medical-journal databases but also summaries of patient records faxable straight to the emergency room, and even face-to-face video encounters between doctor and patient.

Law firms and corporations have just begun collaborating in an online network created by Steven Brill of American Lawyer Media that treats legal briefs and sample contracts like any other database fodder. The idea is that young associates all over the country are duplicating effort, day after day and year after year—at the expense of their clients, of course—re-creating legal research that could more efficiently be standardized as a commodity.

President Clinton's campaign had set up several e-mail addresses even before he took office. On CompuServe try **GO WHITEHOUSE.** On the Internet, **Clinton-info@campaign92. org.** Virtually anyone with a computer can subscribe electronically to the full range of presidential news releases and speeches. "We've gone from a man who didn't recognize a grocery scanner to a new President and Vice President who not only know what PBX stands for, they know the capacity of the one at the White House, and they don't like it," says Representative Edward J. Markey, chairman of the Subcommittee on Telecommunications and Finance.

Executive departments are already online, in bits and pieces. Congress is going online, slowly. The Library of Congress is trying to make its 100-million-item collection—books, maps, scores, photographs—more easily available in the networked world, and few possibilities are scarier than that to existing commercial information services, whose databases are expensive and carefully guarded.

It's true that the culture of the Internet and other parts of the online world reflects a popular feeling that if you've got some text and it's interesting, you should just go ahead and e-mail it. Part of the problem is a psychology born in cloistered college campuses, says Steven Metalitz of the Information Industry Association, a trade association for electronic publishers and database companies. "The idea that you can send this stuff around on your campus and have it covered by the umbrella of fair use is outdated, because you can now send it around the world as easily."

For all that information swimming through the ether, the crucial measurement is always bandwidth: information-carrying capacity, measured, typically, in bits per second. A standard phone conversation uses on the order of 10,000 bits per second. To send a fax—a picture of a page—over that standard phone line takes a long time because there are so many more bits of data in a picture than in a page of plain text. The little television picture in one of AT&T's new Videophones starts out at about 10 million bits per second—so the information has to be compressed a thousand times before it can squeeze through an ordinary phone connection.

There is something misleading about the metaphor of an information "superhighway" to carry the ever-growing volume of data across the nation. At the superhighway level, most telecommunications experts agree that there is no shortage of capacity, with high-bandwidth fiber-optic cables being laid as fast as they are needed. The entire volume of Internet traffic across the Mississippi River amounts to a single 47-megabit fiber channel—a thousandth of the total telephone-network volume. Where data reaches a choke point is in the last half-mile of the journey—from the local phone company central office to the home. There it travels over the twisted-pair copper wire that has carried phone conversations for a century. An ocean of data is sloshing around out there, and most of us are trying to sip it through a very narrow straw.

The coaxial cable that cable television companies have brought to most houses has a capacity hundreds of times greater, and some cable television companies are already going even further and laying fiber-optic cable. Local phone companies have begun to worry about the competition sure to come from the cable television companies, with their large bandwidth in place. The telephone companies have not even begun the vastly expensive process of replacing that last half-mile of copper, but it is possible to push much more data through standard phone wiring. Ordinary wiring can be made to carry the equivalent of hundreds of phone calls at once, or many-page faxes in seconds rather than minutes, or high-quality video images. The method closest to real availability for most consumers is ISDN (Integrated Services

Digital Network), which increases bandwidth by a factor of about fifteen, to almost 150,000 bits per second.

ISDN has become almost a crusade for telecommunications fans. It is a major priority of the Electronic Frontier Foundation, an organization founded in 1990 by the software pioneer Mitchell Kapor, dedicated to the idea that electronic communication is creating forms of community that were not possible before and that, as the foundation grandly puts it, "these communities without a single, fixed geographical location comprise the first settlements on an electronic frontier." Local telephone companies have been notoriously slow in offering ISDN to consumers, but a national standard has finally been set, and it will be available in at least portions of most cities by the end of this year.

An even more advanced digital technology, DSL (Digital Subscriber Line), promises another jump in the bandwidth available in that old copper wire. And in the ground, it is at least always possible to lay another cable. Not so in the air. The electromagnetic spectrum is not so infinitely extensible, and in fact it has long since mainly been spoken for. Since it was decided in the prehistoric first half of the twentieth century that the portion of the spectrum from 54 to 60 megahertz belonged to television Channel 2, that portion has been gone. To retrieve it for use in, say, pagers would inconvenience the owners of the nation's television sets and television stations.

Spectrum is a scarce resource, and particularly in the United States, which has led the world so long in making use of what there is. Major cities, beginning with New York and Los Angeles, have the worst spectrum crowding. The Federal Communications Commission is struggling to reallocate a small portion for personal communications uses, and every allocation decision is fraught with pressure from competing interests. Should a few megahertz near the Channel 13 range stay with its moribund maritime uses or be taken over for wireless home shopping and banking? Here and there the FCC asks police departments and utilities to give up—or sell—their rights to spectrum. President Clinton is supporting a large-scale auction of spectrum, expected to raise $4 billion for the federal government; small companies contend that such giant auctions will shut entrepreneurs out of not-yet-mature markets. One way or another, the government is headed toward a reallocation of 200 megahertz, the first significant transfer in a quarter-century.

The essence of the spectrum problem is that technologies like pagers and cellular phones are trying to use the airwaves in a way that defies modern communications history. Traditionally, the wired telephone network has been used for point-to-point communication, while radio has been used mainly for broadcasting to dispersed groups. The distinction is crumbling. Now faxes can be broadcast to hundreds or thousands of people, while cellular and pager networks work on clever ways to send their signals to individual targets. Switching without switches is a problem. For more than a century the network has operated on a straightforward premise: You place a call from one point to another. The destination is an object fixed in space. As the phone number you dialed flows through the switching equipment, it is parsed one piece at a time: **1**—it's a long-distance call; **212**—send it through to New York; **242**—the Chelsea neighborhood . . . Now the network has to learn to hit moving targets in the telephone landscape. Many phone companies are trying to offer versions of a personal, lifetime phone number that will be yours no matter where you are. It is not easy, technologically—a switch will take much longer to look up a ten-digit number in a vast and ever-changing database than to look up a three-digit area code.

Meanwhile, in Portland, Oregon, the taped voice of Cindy Thompson tells callers, "Leave me a message and my receptor will be notified . . ." A moment later, wherever she is, her wristwatch chirps. When she glances at the dial, she will see your phone number or a coded message or the closing Dow Jones Industrial Average or a Trail Blazer basketball score she has been waiting for. Her company, Seiko Telecommunication Systems, began a trial in Portland and Seattle last year and plans to expand it to larger cities. Subscribers' watches give them weather forecasts, local ski conditions, and lottery numbers. Seiko is working on sending personalized bank balances every morning. Customers are asking for fishing and windsurfing reports.

Michael Park has been struggling to get the wristwatch project off the ground for almost a decade; Seiko took it over after an earlier venture went bankrupt. The technology is ingenious. Instead of using standard pager frequencies, it uses an FM subcarrier band—a slice of free spectrum, next to the main broadcast frequencies, that can be licensed from a radio station. "There's lots of room in the subcarrier," Park says, "and lots of FM stations." The receiver inside the watch is off most of the time, but it wakes up once every two minutes for just a few milliseconds to check for a message. The trick is that, within the two-minute period, everyone's watch wakes up for a *different* few milliseconds. Thousands of tiny time slots are available. Thus a computer-controlled broadcast sends a constant stream of messages to thousands of watches, each popping up at just the right time to catch its own data. Park says this system can handle 300,000 to 400,000 users without being saturated, far more than standard paging systems.

Battery life is a year. Seiko has added a pocket receptor designed to look like a makeup compact—"for women who find

47

the watch too bulky," Thompson says. Anyway, like everything else, next year's watch will be much, much thinner.

It is a peculiar fact of the telephone business that human beings have had trouble mastering the twentieth-century technique known as the Switch Hook Flash.

Engineers thought it would be simple enough: press down the switch hook (the button or lever on which the handset sits when the phone is hung up) for an interval specified by the technical manuals—namely, no less than 300 milliseconds and no more than 1,100. You must execute the Switch Hook Flash to use many of the special services offered by the network. You do it to switch calls when you get the Call Waiting beep, for example, or to transfer a call within an office—which is why you often hear yourself saying, "And in case I lose you . . ."

It turns out that people either press too quickly, in which case the network thinks it has heard some random noise, or too slowly, in which case the network thinks you have hung up. The fancier new telephones have a button labeled Flash—which might remove the element of human error, except that phone company research shows that most people never figure out what that button is for.

The problem of the Switch Hook Flash is symptomatic of how awkwardly the phone companies have tried to graft new services and features on to a technology whose backbone—signals traveling in raw electrical form down copper wires—has not changed in a century. Our telephones are astoundingly stupid, no matter how many buttons and LED's they have. They can communicate with the central office only by interrupting the circuit for a few milliseconds or by sending more noises—touch tones or dial clicks. Caller ID is crude by the standards of computer communication: it works by sending a little burst of noise down the line to the phone in the interval between the first and second rings. If you pick up the phone too soon, you won't see who is calling. And if you happen to have Call Waiting, Caller ID will not work with it, at least for now. No wonder so much of the technology seems awkward and ungraspable. The network is poised to deliver vast intelligence in a digital stream, but these are dumb buckets at the end of the pipeline.

One of the dirty little secrets of telephone marketing over the last decade has been that the many new services offered—Call Forwarding, Call Blocking, Speed Dialing, and more than a dozen others—have been wildly unpopular. About a third of phone users have bought Call Waiting, by far the most successful. The others lag far behind, and the difficulty of using them is surely much of the reason. Not only do users have to master the Switch Hook Flash; they also have to use three-key codes that are hard to learn and remember. If you have Call Waiting, did you know you could block it for the duration of a call by pressing a code like (depending where you live) *70? Now that you know, do you think you will still know it tomorrow?

"Telephone engineers don't worry about that—they say, let the user figure it out," says Barry Schwartz, executive director of voice services at Bellcore. His organization brought in a group of New Jerseyans and sat them down with telephones and standard phone company instructions for adding a third party to a telephone call. Only about 40 percent were able to complete the maneuver.

Technologies like ISDN will remove some of the cobwebs

from the process: the signals that control the progress of a call are separated completely from the voice channel, so they can be handled automatically and intelligently by smart telephones. Even without ISDN, Schwartz has set up a variety of telephones with little screens, like the screens on automated street-corner cash machines, to let users see their choices at any given moment. They are a reminder that integrating our machines—telephones, computers, televisions—will not be nearly as easy as integrating the data in packets of bits on the network.

Meanwhile, the Electronic Receptionist, also under development at Bellcore, will be screening phone calls—presumably sparing us those awkward moments spent hovering over our answering machines deciding whether to pick up the call. The Electronic Receptionist's designers are paying careful attention to etiquette and protocol—touchy areas in the phone business. They want you, when you call Fred, to hear a voice say, "Please wait while I try to find Fred"; they want Fred to hear, "It's Wilma, calling from her car"; and if Fred is too busy for Wilma right now, they want Wilma to be rerouted to, say, a voice mailbox, without ever knowing for sure whether Fred was reached. Human receptionists know how to handle such subtleties, but in many companies, human handlers of telephone calls have been a particularly hard-hit employment category over the past decade, devastated by a rapid changeover to voice-mail and Touch-Tone response systems.

"You have to be very careful what you do," Glenn Eaton says, "when you start to have callers handled by a nonlive person, or a nonhuman." The terminology is still somewhat fluid, but we know what he means. A startling and fast-growing percentage of our calls to businesses seem to involve someone *nonlive* at the other end. Eaton markets systems meant to keep callers happy and "hooked in" even when they never reach a human being. "You don't want to start to alienate them," he says, "or have them feel they're in"—to use the industry term—"voice-mail jail."

There is a fundamental problem with this odd new form of communication, an accidental offshoot of the invention of the Touch-Tone phone. Our brains are just not designed for it. "You can't present complex choice information to people in the auditory channel," says Barry Schwartz at Bellcore. At Bell Labs, Greg Blonder, a research director, does not even use his own company's voice-mail system. "What I *want* to do is pick up the phone and say, 'Get me the message that Bob sent me somewhere last week.'" The network isn't ready for that—instead he has to navigate a branching menu. "I can't retrieve the information, so I don't save it. It's hopeless."

Still, an entire industry has emerged. "There's an infrastructure of publications, and there's boodles of case histories, and there are conferences," says Gordon MacPherson of the Incoming Calls Management Institute. Surely there are limits to what is possible? "It is hard to imagine a machine handling a complaint call," he says. "People communicate with *emotions.*" Still . . . why not speculate? "I suppose in the future we could have devices on the lines that detect the caller's stress level and, based on that, access a prerecorded library of celebrity personalities designed to achieve maximum rapport and manipulate the hell out of us."

Most American households will soon receive, if they haven't already, a notice of a small change in the nature of the phone service available to them. Each time the phone rings, they may see the number of the caller. In return, they should be aware that their phone numbers, too, will be available to the people they call.

More than any other piece of the telephonic future, the introduction of what is usually called Caller ID has alerted users to profound changes coming in the nature of privacy. It has set off a debate: women's advocates and others have filed lawsuits trying to stop or compromise Caller ID. If any one image is motivating the opposition, it may be the image of a battered woman calling her abusive spouse (to demand overdue child support, say) and helplessly revealing the number of a shelter—"thereby endangering," as one lawsuit puts it, "all current and future residents." The phone companies have been largely taken by surprise; they did not expect the right to conceal one's phone number to be as deeply cherished as it has proved to be.

We reserve the right to peek out before opening our door to a stranger, they argue. Why not before answering the phone? Suppose Caller ID had been a feature of phone service from the beginning. After all, it is only a technological accident that we have *not* been able to see callers' numbers. If the phone companies now offered a new service that would enable people to conceal their numbers while making a call, would that not outrage privacy advocates? Wouldn't it open the way to all sorts of intrusions: harassing and obscene calls, for example?

It is not an easy issue. However, in one way the battle was over before it began. Few people realize it, because the phone companies have been notably circumspect, but any business that uses an 800 or 900 number can *already* see and record—in real time—the name, phone number, and billing address of every caller. This business service, called Automatic Number Identification (ANI), works far more effectively than Caller ID. For one thing, Caller ID does not yet handle long-distance calls. And with ANI, the caller cannot block the identification. "There is a major philosophical difference," says Daisy Ottmann of AT&T. "We're giving it to the people who are *paying for the call.* (Well, all right— "we're not actually giving it, we're selling it.")

Thus, a business you call can look up your phone number in its computer database before it even picks up the phone. Many mail-order houses already do this, though they tend to be tactful so as not to spook callers. "It doesn't make any sense to say, 'Hi, Jim, I see you're calling again, do you want to order another smoked turkey,'" Ottmann says, "but it certainly saves time if your address doesn't have to be retyped." Businesses are just beginning to see the range of potential applications. Brokerage firms can use caller identification for security, ensuring that callers are who they say they are. A power company can divide calls by geographic region and offer customized information during blackouts. The possibilities are as varied as they are unnerving.

AT&T says it discourages business customers from using this service to compile and sell phone-number lists on the model of mailing lists, but if that's what a business wants to do, no law prevents it, despite various proposals pending in Congress. To whom, you might wonder, will a dial-porn service wish to sell its customer phone list? Phone number databases are a major commodity. One list compiled for the 900-number industry—extra-cost services from phone sex to newspaper crossword-puzzle answers—tracks people who have refused to pay, legally or not. A 900 service can automatically match a caller's number with the database and hang up.

Ultimately, these emotion-tinged arguments will recede into

a more complex background. Far more difficult and subtle issues affect the future of privacy in a world ever more connected. It may or may not give comfort to privacy advocates to learn that the institution most actively worried about recent developments is the federal government. Law-enforcement agencies have realized that the coming of digital networks seriously compromises their ability to eavesdrop on phone conversations.

When calls are digitized, compressed, and intelligently intermixed with other data on their way down ever more capacious cables, it becomes impossible for an agent to attach a pair of alligator clips to the wires in your basement and simply listen in. The Clinton administration has now proposed a national standard for encryption chips to be built into all sensitive government phone equipment and networks, as well as commercial devices to be sold to businesses and individuals. The plan combines a new and advanced encoding scheme with a catch: the keys to the code will be retained so the phone equipment will be "wiretap ready" for law-enforcement or national-security eavesdropping. Only a government contractor may manufacture the chips, and only the National Security Agency, the highly secretive body that designed the plan, knows the details of the algorithm.

In theory, other encryption devices, using non-government standards, could still be used by any pair of spies or gangsters hoping to evade federal eavesdroppers. That would make the agency's plan useless exactly where it's most needed. So the government hopes to discourage the sale of alternative encryption devices— and AT&T now says it will switch exclusively to the government's chip for its "Bump in a Cord." At the extreme, the plan raises the nightmarish vision of a nation where authorities can monitor all telephone traffic at will. Representative Markey of the Subcommittee on Telecommunications and Finance has raised questions of both privacy and practicality. "It could give them much more information than would otherwise be available," Markey says. "We can't allow them to just wander around in the cyberspace world."

Meanwhile, whatever the government may know about us, it seems that the network itself—that ever-growing complex of connections and computers—will know more. And no matter how much we may bristle at that idea, we nevertheless seem to want services that the network can provide only if it knows.

"There's a balance," says Greg Blonder at Bell Labs. "When you're using a cellular phone, do you want the network to know where you are? You might want the network to know, because maybe you want it to automatically warn your secretary you'll be late and give an estimated time of arrival based on your average speed, or help you get around a traffic jam. But you may also want to hide."

Blonder's office is filled with toys: phones with different coatings—sandy or suede—and electronic gadgets. He holds in his palm a device that reads radio signals from satellites and triangulates his position within 100 feet or so. (It would be even more accurate, but the United States government, for military reasons, requires the signals to be kept slightly fuzzy.) If you turn it on when you leave your car in the shopping-mall garage, it will monitor your path and guide you back—no more balls of thread to get out of labyrinths. Any consumer can buy one for about $1,000; in a few years it may cost a few hundred dollars, and a few years after that, it may be built invisibly into our pocket telephones.

"A human desire that needs to be catered to is the desire to know where you are—not to be lost," Blonder says. "You're going to see a lot of hardware and services using global position-ing. But there's an issue here. That means that marketers can figure out where you walk every day. Maybe they'll use that information to find the best place to site a coffee shop. Now, that's good and bad, right? Maybe you consider that a service—they put a coffee shop on your route. On the other hand, maybe you don't want someone else to know that you're always taking that trip at lunch."

Network-linked kiosks are also spreading, on the model of bank teller machines. Kiosks now offer chocolates, flowers, and Federal Express air bills. They are all connected, and, in theory at least, the information tied up in each transaction can flow both ways. As long as we want to be fed, rescued, or informed by an intelligent network, we will compromise our privacy. Or if we merely wish to remain . . . in touch.

If it all seems mildly Orwellian, a research group at Bellcore has gone even further, embracing the ultimate Big Brother technology: the network that watches *you*. For two years, more and more employees have signed up for an experiment called Cruiser, which—if everyone weren't so cheerful about it—would surely seem like the final invasion of privacy. There are now 140 people with tiny television cameras resting atop com-puter screens or bookshelves. Every so often a soft chime alerts them that someone is watching, and the image of the caller appears in a window on their screen. They may chat informally for a few minutes or interrupt to start a call or continue a call with someone else in yet another window. It seems second nature to hold an impromptu conference with a small group. Cruiser is a world apart from the videophones that have made a tentative reappearance in the consumer market in the past year, decades after the doomed Picture Phone gave AT&T a notorious and painful marketing disaster.

The new generation of videophones represents a small victory of technology over the difficult constraints of existing telephone wires. With ISDN or fiber-optic links, far clearer pictures will be possible, as will widely touted new possibilities for entertainment: libraries of movies to be shot down the phone lines on demand; television broadcasts that let the viewer serve as the director, panning across the football field or zooming in on the action as he sees fit. Meanwhile, the latest videophone efforts by AT&T and others compromise by offering a tiny, herky-jerky, but fairly sharp image. Consumer research showed that people were somewhat less irritated by jerkiness than by fuzziness.

"If all you were doing in the communication act was conveying information, then visual doesn't add a lot," says Robert Fish at Bellcore, a cognitive psychologist who is one of the project's designers. "But that's a naïve view of communication." People drop in on each other more easily than they phone; they express moods, subtexts, emotions. Of course, those are *information,* too. "We've done studies showing that most communication in business is informal, unscheduled, face-to-face—an ancient technology called proximity."

Workplace sociologists know that when people leave the office to work at home, or even move to a new floor just an elevator ride away, their sense of belonging drops rapidly. They fall out of the loop. Cruiser is meant to replace not formal conferences but the unquantifiable water-cooler and stairwell meetings. It seems easy somehow. Ordinary phone calls between Bellcore employees linked by Cruiser tend to drop from five or six a day to zero.

At Bellcore participants can easily turn Cruiser off or set it so they cannot be seen until they answer a call. But most find themselves leaving it on and allowing callers simply to drop by. It is easy to imagine less friendly uses of the system: supervisors could use it to monitor employees, for example. In Fish's view the technology itself is neutral. If a workplace is open and friendly, then Cruiser is open and friendly. If not . . .

Fish brings his prototype to a halt in the course of trying out some advanced features, and at once the phone rings, the old-fashioned way. A telephone that needs to be picked up and placed to his ear seems, suddenly, an intrusion.

At any given moment, several dozen people in the United States, Britain, Taiwan, Hong Kong, Norway, Sweden, and a handful of other countries can be found—if *found* is the right word—playing bridge with one another across the Internet. It is a bizarre and yet utterly characteristic Internet activity. Tiny amounts of data are involved: to relay the bids and the play of the cards consumes a microscopic percentage of the billions of bits flowing through transatlantic and transpacific cables. One high-resolution fax carries more data than a week of bridge—then again, people are also playing chess, backgammon, Go, and other games. The computer programs that sustain the games are distributed across the network, set up on host machines in thousands of different places. Each bridge player sees a crude representation of a table, with hands dealt automatically, and as the game proceeds, so does an odd conversation, carried out in a sort of post-Pinteresque pidgin, one line of prose at a time across the bottom of the screen.

kvk: I read, I played, I conquered.

denny: is it my imagination, or are there a lot of new players on?

jg: bye. real life calls.

endora: oh, like this isn't?

The same hands are played again and again by different pairs, who compare their scores, so in effect it is a global, round-the-clock, never-ending duplicate bridge tournament. The program and the playing are free for anyone with access to the Internet. Commercial gaming services are also springing up. "Recently I've become aware that computer bulletin boards seem to have become the basis for a new subculture," says the program's author, Matthew Clegg, a graduate student at the University of California at San Diego. "Perhaps one attraction is that it offers a way to meet other people in a completely nonthreatening setting. As strange as it may seem, I know of several people who like the game precisely because it does provide an alternate form of socialization."

So is the network driving us apart or bringing us together? Some scholars fear that it fosters impersonal communication. By tradition, a telephone conversation or a packet of e-mail is human contact drained of nuance, gesture, smell. But social theorists have been predicting a decline in community and a rise in alienation ever since they began to pay attention to the industrial revolution, and it is far from clear that they have been right.

"One of the crazy things that's happened is that we do still form communities," says Scott Patterson, assistant professor of communication studies at Virginia Polytechnic Institute, who has been studying the social consequences of telephone networks.

Our communities are no longer the fraction of a square mile around our homes. We are developing an ability to relocate ourselves within many more slices of the human universe. In the past, people with an obscure or even not-so-obscure special interest—medieval fabrics or military-strategy games or expressionist art—made their way to cities or universities, the only places that accumulated critical masses of like-minded people. Now the possibilities are richer. Patterson is part of an "electronic village" project in Blacksburg, Virginia, one of several under way in different parts of the country, trying to join schools, libraries, town buildings, and homes in intelligent networks. We are learning, Patterson says, to maintain new forms of "virtual community" or "community without propinquity."

"The telephone is a technology that allows us to transcend these things that limit us as human beings," he says. "The thing about the telephone is that it allows us to transcend space but not very well. The answering machine allows you to transcend time a bit. Computer networks allow us to transcend both time and space."

Sometime during the telephone era, science fiction writers began to imagine worlds, or futures, where whole populations would be linked with telepathic efficiency to create a supermind, an organism greater than the sum of its parts. "Each imprisoned in its hollow world, and physically isolated from the rest of the cosmos, these populations telepathically supported the cosmical mind," wrote Olaf Stapledon, an early practitioner of the genre, in 1937. "Intimately they knew one another in all their diversity." Is *that* where we're headed?

Martin E. P. Seligman, professor of psychology at the University of Pennsylvania and an Internet bridge player, has been monitoring the temperature of emotional incidents on the computer network to see how they differ from "real time" emotion. The constraints of the form tend to flatten anger and push toward civility, he finds. He suspects that the electronic classroom may favor different kinds of people. "In real life," he says, "a political, spontaneous animal comes to the fore. A thoughtful, shy person might flourish on an Internet university."

The network filters us, reducing our movements and speech to the tempered voice on a telephone or the plain words on a computer terminal. In return we reconstitute the freeze-dried voices that come to us—we create mental images to flesh them out. After getting to know someone on the network, we can be startled by a real photograph or a real voice. We feel we have been hearing the thoughts themselves, free of superficialities. It is possible for a relationship to survive the ether-to-flesh transition, though. At least one pair of Internet bridge players have married since first meeting online.

The owner of a Melbourne, Australia, grocery store recently discovered that her daughter, a non–bridge player, was surreptitiously using the program late at night to chat up boys on various continents. When Seligman was playing a bridge hand not long ago, he realized that some of the other players and kibitzers belonged to a loose subcommunity of Chinese expatriate students that has formed online, one of many such subcommunities—in this case mostly firstborn sons sent from the mainland to make their way in the world, now looking for an extended family to take the place of the families they miss, Seligman felt. As the cards crossed the screen and the lines of chatter rolled by, he began to tell about a young man he knew who had been dying of cancer in California, far from home. Near the end, the young man brought a portable computer and modem to the hospital, Seligman typed—

so he could die among his friends

The electronic conversation came to a halt. There was a pause and then, through the silence, a sort of stage direction:

alex: [moved]

And another:

rou: [deeply moved]

THE INFORMATION FUTURE
OUT OF CONTROL, FOR GOOD

May 1994

Here I am, late one evening, on the telephone with an elderly man in his home somewhere in Manhattan. I don't know his name. He doesn't know mine. I am frantically begging him to leave his handset off the hook. My Chinese is even worse than his English—hopeless.

I have opened a new outpost in the electronic landscape, a company called the Pipeline, dedicated to offering everyday access to the blooming universal network of networks known as the Internet—aka, Information Superhighway. Our customers are arriving home from work, switching off the evening news, turning their backs on spouses, and trying to dial into our gateway. We have banks of telephone lines feeding into a network of computers—our network, in turn, attached via a series of high-speed digital circuits to the global Internet. Our telephone lines are arranged in standard fashion, so that, as each becomes busy, a new caller is automatically bounced to the next line in the sequence.

For reasons that will never be explained, however, the Nynex central office at West Street has suddenly begun bouncing a line at the very beginning of the sequence to a random residential number elsewhere. The rest of our telephone bank has effectively been cut off. Our victim, meanwhile, has been answering his phone resolutely, minute after minute, greeted each time by a sound human ears were never meant to hear, the squeal of a modem. If he would at least leave his phone off the hook, the Pipeline's customers would be bounced back into our sequence— but why should he? He has paid for his telephone service, too. The Nynex repair number, 611, is as always a black hole, offering a recorded voice that promises to investigate the trouble by the end of the next day.

I have seen the future, and it's still in the future.

The computer, television, and telephone empires have seen the future, too, of course. Never in the annals of business have so many great enterprises raced their wheels so violently in the cause of so seductive a vision: video beamed over telephone wires; telephone calls digitized and packetized over television cable; data and interactivity everywhere. If you want to pan to the third baseman while other viewers are watching the batter, you can— in the laboratories. Want to read the newspaper seconds after it's transmitted to your pocket computer? Surf 500 channels, replay that sex scene, talk back to Connie Chung? Easy. Come and get it!

Information providers and information-provider wannabes are equally frenzied. If you're a major newspaper or news service, or if you're a giant entertainment conglomerate, you are pouring money into pilot projects and online trials. In the confusion, *everyone* is an information provider. It seems that every author, game developer, cartoonist, porn star, and greeting-card designer in America has already signed up for the thing called *multimedia*— a term that is usually a fancy synonym for CD-ROM, laser disks that can enliven your computer with the multiple media of words, sounds, pictures, and snippets of video. The vast majority of computer owners don't even have CD-ROM drives. If they don't act quickly, they will miss the chance to play the Too Many Typefaces CD-ROM, the Fractal Frenzy CD-ROM, the Clinton Health Security Plan (What It Means) CD-ROM, the CIA World Tour, the San Diego Zoo's Animals, the Anglo-Saxons, Learn to Play Guitar . . . Meanwhile, the word in financial circles is that a business card containing the word *interactive* will pass the bearer through any door in corporate America.

Amid the froth is a conviction that nothing less is at stake than the entire future of the world's information, communication, and entertainment infrastructure. If only one could find just the right strategic alliance, just the right corporate merger, just the right software vendor to help that video-music-text-art combination soar through the . . . airwaves? Phone lines? Television cable?

Unfortunately, the present has a way of staying with us, and these great companies are already littering the business landscape with the debris of their shifting strategies. The merger of two of the largest, Tele-Communications Inc. and the Bell Atlantic Corporation—reported, analyzed, and financed on the scale of a new Sino-Soviet alliance—materialized and dematerialized in a blizzard of press clippings. Time Warner Cable has revealed that its field trial of interactive entertainment and home shopping, meant to begin this spring in 4,000 Florida homes, will have to be put off till the end of the year, at best. A regulatory change here, a software-development problem there—and the future recedes once again.

The Information Superhighway has reached buzzword status so fast that no one has time to utter all eight syllables anymore. The latest spin-off coinages are *Infobahn* and, most succinct, *I-way*. "It's showtime on the Infobahn," people keep reminding us. The I-way has on-ramps, potholes, traffic jams. "There are speed limits on the Information Superhighway" is almost as popular a cliché as its opposite, "There are no speed limits on the Information Superhighway."

In fact, there *are* speed limits on the Information Superhighway. If your computer uses a 14,400-baud modem, you are among the lucky elite, but you will find yourself drumming on the table as you wait for today's satellite photo, and online video is out of the question: too much data to squeeze through too narrow a channel. No one has done more to give American business the official I-way go-ahead than Vice President Gore, who understands the cultural and economic power of universal connectivity. But there was no video, no audio, no multimedia in Gore's celebrated first electronic town meeting. There was only Gore, happy technologist and, luckily, touch typist, rattling words out through his computer keyboard to an audience of hundreds. The speed limit was measured in words per minute.

For large institutions like universities and data-processing companies, with huge quantities of information to send back and forth daily, the notion of an Information Superhighway isn't a terrible metaphor. There is an infrastructure that needs to be expanded and maintained: high-capacity electronic paths across the country and around the globe. There are routing problems and traffic problems. There really are tolls and bridges, on-ramps (sort of) and potholes (unquestionably).

For most of us, however, the metaphor is misleading. We don't have continent-sized data blockages waiting for the construction of a giant cross-country conduit. We have ancient copper telephone lines that we wish could somehow support our fax machines, modems, and voice conversations. The choke point is in the few blocks between our homes or offices and the telephone company's switch. The miracle is that so many people and small businesses are managing to find their own way into the electronic world . . . "that electronic world which more and more supplants the dull world of heavy elements and three dimensions," as John Updike put it recently.

It is growing not by design but by accretion. It resembles not a fat, linear highway but a protoplasmic organism, or colony of organisms. New bits are constantly floating along and adjoining themselves to the main mass. When my company adds a new customer—especially a customer who stays online for hours at a time, sending out e-mail and commentary, or placing information online for others to stumble upon—the Internet has grown by another degree.

Somewhere out there, people are taking the trouble to put the professional sports schedules online, free, to be consulted by any surfer of cyberspace who suddenly feels the urge to know where the Indians are playing on May 4.

A pair of architecture buffs have started assembling a "multimedia resource" dedicated to "the dissemination of architectural knowledge": Palladio drawings, Kandinsky paintings, musings on "lunar architecture"—and of course links into everyone else's equally new, experimental, and personal sources of architectural information.

Debbie (the Leaper) Brown, working for a computer company in Rochester, took the trouble to post a complete episode guide for one of her all-time favorite television shows, *Miami Vice*.

Bill Sherman at the National Center for Supercomputing Applications, more of a Muppets fan, put online a "Muppography."

Students at Carnegie Mellon have wired in a soda machine, so that a thirsty Internaut in Paris or Taiwan can get a real-time display in a typically raw style:

```
M&M validity: 9 Coke validity: 9
M & M Buttons
/-----\ C: CCCCC .................
|     | C: CCCCCCC ....D: CCCCC ......
|**** | C: C ...........D: CCCCCCCCC...
|*****| C: CCCCCCCC ...D: CC ..........
|*****| C: CCCCCCCCC ...
\-----/ S: CCC .........
| Key:
| 0 = warm; 9 = 90% cold; C = cold; . = empty
| Leftmost soda/pop will be dispensed next
---^---
```

This is the Storehouse of Human Knowledge, Department of Grass Roots. Nothing is too trivial to find a permanent place: certainly not this week's Nielsen Ratings, and certainly not the Twin Peaks Allusions (sixteen contributors from the United States, Britain, and Sweden), the "Twin Peaks Pilot—every second of it," Twin Peaks Symbolism, or Twin Peaks Timelines.

It's surely in the same spirit that the government of the United States has begun tossing online its own miscellany of useful information. The new Internet site of the Senate itemizes **Available Documents Distributed by Member**, and who cares that so far it's an online Senate of four (Patrick J. Leahy, Edward Kennedy, Charles S. Robb, and Ted Stevens)? The White House posts every position paper and speech and recently set off a debate among citizens who had noticed a nontrivial change ("censorship!") between a pre-delivered and post-delivered version of a Clinton speech. Of course, history is a continuum, and the Federalist Papers are available next door (cyberspatially speaking).

The hardest fact to grasp about the Internet and the I-way is this: It isn't a thing; it isn't an entity; it isn't an organization. No one owns it; no one runs it.

It is simply Everyone's Computers, Connected. It is the network of all networks—the combination of all the large and small university, government, and corporate networks. It extends to individual PCs at the end of the line, like shacks at the ends of dirt roads not far from the turnoff to U.S. Route 1.

It has taken shape with startlingly little planning. It received only the most accidental assistance from top government policy makers ("Information Superhighway" was no more a Bush administration watchword than was "Supermarket Checkout Scanner"). Nor did the telecommunications companies help much—it is their fiber carrying the megabits of data, but they remain conspicuously absent from the business of getting companies and individuals onto the Internet. No, the most universal and indispensable network on the planet somehow burgeoned without so much as a board of directors, never mind a mergers-and-acquisitions department.

There is a paradoxical lesson for today's strategists. In economic terms, the great corporations are acting like Socialist planners, while old-fashioned free-market capitalism blooms at their feet. We seem to live in an era when giant communications empires own the cables and airwaves and giant information empires own everything else—when Simon & Schuster and Warner and Paramount and CBS Records and *Time* magazine and Sony and hundreds of magazines, cable-TV stations and rap-music companies are, if not yet one big company, approximately three. Yet here's the Internet, a world controlled by no one, like a vast television station without programmers or a newspaper without editors—or rather, with millions of programmers and editors. It's a frontier, befitting its origins: unruly, impolite, and anarchic. But also democratic.

My own obsession with the Internet began with sheer wonder at the junkyard plenitude of information, tempered by horror at the difficulty of finding anything. Most people lucky enough to have found a way to dial into the Internet have found themselves confronted by one of the world's strangest linguistic phenomena, the operating system known as Unix. They have also had to know the computer addresses for each item on this giant library shelf. Want an hourly status report from Canada on the activity of the aurora? Just enter the command: **finger aurora@xi.uleth.ca**.

This seems unnecessary. The point of the Pipeline has been to create an environment, with simple, graphical software, that organizes at least some of the wildness. We try to cut pathways into the jungle, even if the underbrush does have a way of growing back. Certainly graphical interfaces are the future, and they are bringing a new population into the electronic world. Is that a good thing? The original inhabitants don't always think so. Part of the Internet's culture, and not the most attractive part, has been a form of elitism that has encouraged the obscurity. It has been like a town that leaves its streets unmarked on the principle that people who don't already know don't belong.

I have visited the advanced telecommunications research laboratories and seen what technology can bring—the digital telephone service called ISDN, for example, which promises to turn ordinary phone lines into high-bandwidth carriers of pictures and videos. Lately, though, on behalf of the Pipeline, I've also visited the local telephone company and seen what technology can't bring. I've tried to order this very service. I have a fourteen-page, four-color brochure! "Nynex ISDN Primary Service . . . for More Efficient Voice, Data, Image, and Video . . ." The Pipeline's ISDN order has been floating about for months. Our sales representative says he wrote it up three times, and each time the system bounced it back. I have a phone number for an ISDN specialist inside Nynex, but he doesn't seem to have voice mail.

Luckily our customers understand the environment we're working with. "Sorry, my response to Dave's post on Indian beach food got eaten up by a second-level demon somewhere in the Giant Tunnel of the Fourth Moon of Nynex," writes a New York financier in one of our discussion forums. We make do. We've discussed plausible advanced-technology scenarios for bringing our necessary high volume of telephone circuits into our office: Nynex has plenty of fiber and packet networks, Enterprise Circuits and Infopath, and Advanced Custom Networks. We don't want hundreds of separate telephone numbers (a scarce commodity) and we don't want dial tone (our customers call us). But we get them whether we want them or not. Somehow, when we file out of the conference rooms, the solution is always the same: a wall full of individual, old-fashioned telephone lines.

The Pipeline is not alone. The large private online services, CompuServe, Prodigy, and America Online, all rely on graying technology—at least we can offer our users the highest available modem speeds. Those services may have a limited future anyway. Either they will open their gates to the Internet and become subsumed by it, or they will remain lakes isolated and apart from the ocean. Meanwhile, despite ourselves, we have become revenue producers for the telephone companies. One of our customers racked up a $544 phone bill in her first month, while paying the Pipeline a total of $35.

Maybe this is the era of small warm-blooded mammals scurrying about at the feet of the dinosaurs. Our reason for being, a graphical software package that guides new users through the complexities of the electronic world, was the half-year's work of a lone, overstressed programmer, Uday Ivatury. The established software companies, from Microsoft down, leak monthly rumors of their own online software in progress, but they have not yet produced any. We haven't tried to propel our users headlong into the twenty-first century. They can't receive video on demand. But they can order flowers from a cyberflorist, and they can have their e-mail forwarded to their pagers. They can't surf the famous 500 channels, but they are trying out a Parisian's interactive guide to his city's Metro, or browsing last week's SEC filings through a free experimental project, or joining the arguments in the alt.tv.melrose-place newsgroup, or logging into the Library of Congress, or trying to download this morning's infrared satellite map—and discovering that the weather archives are overloaded.

There are billions of dollars in search of the future, but it's the present that ordinary users have to cope with. We're amateurs, avowedly; and here on the I-way, it is Amateur Hour.

OPEN, SESAME

April 1995

I had a good password once. I lost it to modern vandals who cracked an Internet node with a password-sniffing program that sat quietly in the shadows and recorded everyone's connections to other sites across the planet. In the aftermath, there was my password, **s!itnol** (memorable as the first letters of the first words of a Supremes song—and you're welcome to it), shockingly exposed near the end of a long security-agency printout. It felt like walking into a gallery and seeing your own nude posture photo. And what a loss! A good password is hard to find.

A bad password is what you are probably using right now, or soon will—for at this peculiar moment in human history we find ourselves obliged to punch in special sequences of characters before we can go online, or get cash from a machine, or check voice mail, or turn off a house alarm, or telephone with a credit card, or reactivate a car radio, or secure a spreadsheet. **Lizzie** and **123184** are screamingly bad passwords, especially if they happen to be the name and birth date of your firstborn. **Gan-**

74

dalf is a bad password; so is any obscenity in any earthly language; so is any name ever used in an episode of *Star Trek*. It's no good switching to **fladnaG** or **481321** or **eizziL**—password crackers do use computers, you know.

The online message groups are hopping with queries like this one from a gentleman in Kuwait:

> **hi there. I want a program which can get others PassWord. E-mail me Please. Nassib.**

Open, Sesame just won't cut it anymore, eh, Nassib? Passwords have become defining tokens of our electronic age—no longer the property of sentries and spies. We've reached a level of interactive networked existence where faceless human contact is the rule, and every connection requires the magic word, not to get the treasure or enter a speakeasy, but just to take the first step: validate our puny existence; prove we are who we say we are.

Password technology now mixes the archaic and the hypermodern. Examination of password files has shown that most people really do use names and birth dates of spouses and children. The customer-service departments of bank-card and telephone-card companies have had to gear up to handle the flood of calls, thousands daily, from customers who forget their Personal Identification Numbers. It is a strain. Security experts at Carnegie Mellon University estimated last year that more than a million passwords had already been stolen on the Internet. Their advice: change your password. Change it again. Naturally, sniffers have learned to watch specifically for the change-password command.

The arms race escalates: the authors of password-cracking programs war with (though they are sometimes one and the same as) the authors of password-securing programs, meant to save you from yourself. One typical tool of system administrators wags a stern finger if you try to create a *bad* password:

too short;

too close to your last password;

a date, a telephone number, or a Social Security number;

a name, a dictionary word, or a name or word spelled backwards;

a two-word combination;

a naughty word (defined in one textbook as any substring of **ibmdecsunat&tnasa**);

an alphabet sequence (**stuvwx**);

a repeating pattern;

a keyboard sequence (**qwertyu**);

a palindrome;

or any entry in an ever-expanding list of Standard Passwords That Thousands of Users Think They Just Dreamed Up for the First Time—**heinlein, lothlorien, popeye, ilbcnu** (I'll be seeing you).

Think you know an obscure foreign word, do you? An Internet site at Oxford University maintains for this purpose lists of millions of words and names in sources from Afrikaans to Esperanto.

So rule out anything you could possibly hope to remember. On the other hand, the worst password of all is one that you've written down on a piece of paper or on the inside of your desk drawer.

It's both good and bad that password creativity has come so

far since **"Who's there?"** **"Nay, answer me. Stand and unfold yourself."** **"Long live the King."** A password is more than just a flaky kind of fingerprint. We still want passwords to be romantic, not just utilitarian. We reveal ourselves in our passwords. That may be one reason it hurts to lose a good one. Passwords are about identity, after all. Choosing **xerxes** or **donjuan** may be a grown-up equivalent of wearing Power Ranger underwear.

Now, let's say you decided to be clever and use password security on your new PC, or you chose a highly creative PIN for a telephone charge card you seldom use, or you password-protected that WordPerfect listing of your collection of pornographic videos—and now you're locked out. Either your memory has failed you or the technology has hiccupped. What next?

You could buy commercial software that defeats the password protection built into other commercial software. One manufacturer in this hot category, CRAK Software of Phoenix, says its exceedingly grateful customer base includes not just bad guys but FBI agents, lawyers, and internal company auditors and managers trying to recover from a new kind of low-grade sabotage: "password amnesia" in disgruntled and dismissed employees.

What is a human of normal memory capacity to do? The subculture has evolved coping strategies. First letters of not-too-common phrases or lyrics can be good, though experts have been known to claim that techies all listen to the same songs. One professional started with **nuts2u** (none too secure, really) and let it morph: **Nuts4U** . . . **nuTS8u** . . . Another, with responsibility to invent scores of ever-changing passwords, draws from the namespace of esoteric foods, distorted with punctuation marks and digits: **d1M.$um, $paeTZle, KØchoRee** . . .

Technology is lurking, ready to rescue us with smart cards

and cryptographic schemes. A major bank employs a hybrid of five-digit user PINs and random passwords generated by little electronic "hardware tokens." A hybrid is hard to steal. It's also expensive and a nuisance when you get to work and discover that you've left your hardware token on your night table. Some online services are experimenting with a promising technique from cryptography, known as Kerberos. It foils sniffers by preventing the transmission of clear-text passwords. You get a new one-time password each time you log in. Each session begins with a quick two-way trade of encrypted messages—to let you prove (in one of the apparent paradoxes that flourish in modern cryptography) that you have a secret key, without actually sending the key itself.

Most of us aren't quite ready to be rescued. We continue to cope with what is politely termed "the human factor." Just last month the people at Intuit, the leading maker of personal financial software, noticed that they had shipped out a password that let any customer gain access to their network and read other customers' tax returns. It's going to be a long time till we equip every computer, every telephone, and every shard of electronics with a palm reader or retinal scanner. We're stuck with identifying people by the all-too-fleeting contents of their heads.

THIS IS SEX?

June 1995

At first glance, there's a lot of sex on the Internet. Or, not at *first* glance—nobody can find anything on the Internet at first glance. But if you have time on your hands, are comfortable with computing, and possess an unflagging curiosity about sex—in other words, if you're a teenager—you may think you've suddenly landed in pornography heaven. Nude pictures! Foul language! Amazingly weird bathroom humor! No wonder the Christian Coalition thinks the Internet is turning into a Red Light District. There's even a **Red Light District** World Wide Web page.

So we explore. Some sites make you promise you're an adult. (Fine: you promise.) Let's try **Girls,** a link leading to a computer at the University of Bordeaux, France. The message flashes back: Document Contains No Data. **Girls** at Funet, Finland, seems to offer lots of pictures (Dolly Parton! Ivana Trump!)— Connect Timed Out. **Girls**, courtesy of Liberac University of Technology, Czech Republic, does finally, painfully, deliver itself of a 112,696-byte image of Madchen Amick—you could watch it spread across your screen, pixel by tantalizing pixel, but instead you go have lunch during the download, and when you return, there she is, black-and-white and wearing . . . clothes.

These pictures, by the way, are obviously scanned from magazines. And magazines are the ideal medium for them. The battle cry of the online voyeur is **Host Contacted—Waiting for Reply**.

With old Internet technology, retrieving any graphic image and then viewing it on a PC at home could be laborious. New Internet technology, such as browsers for the Web, makes all this easier, though it still takes minutes for the typical picture to squeeze its way through your modem. Meanwhile, though, ease of use has killed off the typical purveyor of dirty pictures, capable of serving hundreds of users a day but uninterested in handling hundreds of thousands. The Conservatoire National des Arts et Métiers has turned off its **Femmes femmes femmes je vous aime** Web page. The good news for erotica fans is that users are redirected to a new site where "**You can find naked women, including topless and total nudity**"; the bad news is that this new site is the Louvre.

The Internet does offer access to hundreds of sex newsgroups. They're easy to find—in the newsgroup hierarchy **alt.sex** comes right after **alt.sewing**. And yes, **alt.sex** is busier than **alt.sewing**. But quite a few of these groups turn out to be sham and self-parody. Look at **alt.sex.fish**—practically nothing. **Alt.sex.bestiality**—aha! just what Jesse Helms fears most—gives way to **alt.sex.bestiality.hamster.ducttape**, and fascinating as this sounds, it's empty, presumably the vestige of a short-lived joke. **Alt.sex.bondage.particle-physics**, **alt.sex.sheep.baaa.baaa.baaa.moo**—help!

Still, if you look hard enough, there is grotesque stuff available. If pornography doesn't bother you, your stomach may still be curdled by the vulgar commentary and clinical how-to's in the militia and gun newsgroups. Your local newsstand is a far more user-friendly source of obscenity than the online world, but it's also true that, if you work at it, you can find plenty online that will disgust you, and possibly even disgust your children.

This is the justification for an effort in Congress to give the federal government tools to control, for the first time, the content available on the Internet. The Communications Decency Act, now making its way through Congress, aims to transform the obscene-phone-call laws—"Obscene or Harassing Use of Telecommunications Facilities"—into a vehicle for prosecuting any Internet user, bulletin-board operator, or online service that knowingly makes obscene material available.

As originally written, the bill would not only have made it a crime to write lewd e-mail to your lover, it would have made it a crime for your Internet provider to transmit the data packets. After a round of lobbying from the large online services, the bill's authors have added "defenses" that could exempt mere unwitting carriers of data, and they say it is children, not consenting adults, they aim to protect. Nevertheless, the result is a historically far-reaching attempt at censorship on a national scale.

The Senate authors of this language do not use e-mail themselves, or browse the Web, or chat in newsgroups, and their legislation reflects a mental picture of the online world that does not match the reality. The existing models for federal regulation of otherwise protected speech—for example, censorship of broadcast television and prohibition of harassing telephone calls—come from a world that is already vanishing over the horizon. There aren't three big television networks now, serving a unified mass market; there are thousands of television broadcasters serving ever narrower special interests. And on the Internet, the number of broadcasters is rapidly approaching the number of users: uncountable.

Already most live sources of erotic images seem to be overseas. The sad reality for federal authorities is that they cannot cut those off without forcing the middlemen—online services in the United States—to do the work of censorship, and that work is a practical impossibility. Any teenager with an account on Prodigy can use its new Web browser to search for the word **pornography** and click his way to **Femmes femmes femmes** (but it's the Louvre again; oh, well, better luck next time).

Policing discussion groups presents the would-be censor with an even more hopeless set of choices. A typical Internet provider carries more than 10,000 groups. More than 20 million new words flow across the transom every day. The technology of these discussion groups is hard to fathom at first. They are utterly decentralized. Every new message begins on one person's computer and propagates outward in waves, like a chain letter that eventually reaches every mailbox in the world. Legislators would like to cut off a group like **alt.sex.bondage.particle-physics** at the source, or at its home—but it has no source and no home, or rather, it has as many homes as there are computers carrying newsgroups.

This is the town-square speech the First Amendment was for—often rancorous, sometimes harsh, and occasionally obscene.

Voices just carry farther now. The world did not used to be this global and this intimate, at once. Even seasoned Internet users sometimes forget that, lurking just behind the dozen visible participants in an out-of-the-way newsgroup, tens of millions of potential readers can examine every word they post.

So if a handful of people wish to share their private experiences with like-minded people in **alt.sex.fetish.hair**, they can do so, efficiently—the most fervent wishes of Congress notwithstanding—and for better or worse, they'll have to learn that children can listen in. Meanwhile, if a mob of gun-wielding extremists wish to discuss the vulnerable points in the anatomy of FBI agents, they, too, can do so. At least the rest of us can listen in on them, too. Perhaps that is a grain of consolation—instead of censorship, exposure to the light. Anyway, the only real alternative now would be to unwire the Information Superhighway altogether.

WASHINGTON UNPLUGGED

August 1995

If you have time on your hands and cash to spare, you can walk into the Patent and Trademark Office in Arlington, Virginia, sit down at a computer terminal, and browse through the entire technological history of the United States, invention by invention—the drawings and descriptions of Morse, Whitney, and Bell; the paper clip, the transistor, and the accelerating rush of scientific creativity in modern times.

"We have basically all the images and text going back to 1790," says Commissioner Bruce Lehman. "That's unique in the world. This is a fabulous database." It is a treasure trove for scientists, historians, and students—and above all an essential resource for any scientist who needs to see the art and thinking of the inventors who came before.

Now the bad news. It will cost you $40 an hour, a prohibitive price for any but the most specialized uses. If Arlington isn't around the corner from you, you can try some of the patent office's satellite libraries. Alternatively, you can dial into a private data service like Lexis or Dialog and pay even more—fees that can amount to hundreds of dollars an hour, for public information.

This is the information age, Washington-style. As a citizen of the United States, you have accumulated valuable property, in the not-so-ethereal form of data. It's worth billions. And most of it is locked away where you cannot get at it.

Every time an inventor receives a patent, the secrets of the invention become available to the public for their perusal. The same is true of every planetary image beamed from a space vehicle, every corporate filing before the Securities and Exchange Commission, every decision of every judge, every Geological Survey map, every federal health study, and ultimately every memo of the president, scrawled or e-mailed. They all join a virtual public library—arguably the most valuable storehouse of data on the planet. It is also one of the most archaic: scattered across departments, poorly catalogued, and expensive to access. With few exceptions, it is off-line.

Meanwhile, back across the Potomac from the patent office, in a tiny room in the District of Columbia sits the Internet Multicasting Service, not much more than a fast telephone line attached to a donated workstation and disk drives. The service has obtained the patent data for last year and the first months of this year and put it online, free, fully indexed and searchable through easy-to-use natural language queries. Carl Malamud, the service's founder, says he could just as easily offer the entire historical database of the patent office—but the office won't give him the data.

"They treat this as a product, but it's not a product—it's enabling information," he says. It is, or should be, fuel for the information economy.

The patent office already has a high-bandwidth Internet connection. It could easily allow any of the millions of home and business computers with access to the Internet to plug into its system and see just what a user sees at that Arlington terminal, just as any computer on the Internet can now plug into the New York Public Library's online catalogue or the databases of thousands of other libraries. The money has already been spent—more than $400 million to create a patent database available only to walk-in traffic.

So why not go online? The commissioner offers multiple and self-contradictory answers to that question. They echo the reasoning of scores of other government agencies, federal and local, facing the same issue:

It's not our job. "We're not a library," Lehman says. "It's not the fundamental purpose of the patent office. Now if Congress wants to change that they can, and they can provide us with a tax appropriation to do that."

We're doing it anyway, as fast as we can. "It has always been part of our plan to provide a plug into the patent office to outsiders," Lehman says. "We are not at this moment in time in a position to open up that plug for technological reasons."

And, we must not compete with the private sector. "We're developing a big information industry in the United States. We already see about thirty companies that feed off the patent office, and we want to encourage that. Part of what we're trying to do is bootstrap new industries."

That last argument sounds attractive, until you realize that those companies are lobbying for the privilege of paying the government *more*—in other words, they want to forestall competition. They belong to an industry that has used heavy, targeted campaign contributions to protect its stake in an economic model that is rapidly becoming obsolete: scarce data sold to specialists at high prices. West Publishing, with a near monopoly on the government's court databases, is a costly example, as lawyers quickly discover. The Internet has created a different model: information of all kinds, a mass audience, low prices.

Lehman acknowledges that private information services lobby him hard to keep prices up; he denies being influenced by their pleas. Nevertheless, the patent office, like many other federal agencies, sells its data mostly on old-style mainframe computer tapes, at prices low enough to guarantee enormous profits for commercial services but just high enough to prevent widespread low-cost distribution on the Internet.

A potentially far-reaching new statute, passed with little fanfare this spring, requires government agencies to make electronic data available for no more than the "cost of dissemination." Twenty-five years of patent data—for which the patent office charges nearly $200,000—would fit on modern tape cartridges costing no more than a few hundred dollars. How to explain that gap? The office's arithmetic counts not just the tapes and the few technician-hours that would be needed to fill them with data, but also fuzzier items: general staff time, updating the databases, and "customer service."

This is an administration that has jawboned hard for the online world since taking office. The White House has an elegant home page, as does Congress, along with scores of federal departments. They are online with digital renditions of their official seals, their speeches, and their press releases—political information sitting in for the real thing. Yet where the truly valuable databases are concerned, the Clinton administration has produced no comprehensive plan for the future.

It may not be necessary. The White House could take a powerful step forward merely by leaning on its bureaucrats: ordering them not just to comply with the new public-information law but to embrace it. That would mean taking the crucial phrase *cost of dissemination* literally—cheap disks or tapes; duplicating and shipping, and nothing else.

"People are concerned about universal access—the wire running into your house will be the easy part," says James Love, director of the Taxpayer Assets Project, a Washington-based advocacy group. "Certainly the one thing people shouldn't have to worry about is government information, the thing they own as taxpayers. There'll be lots of other things they won't be able to afford. At least this should be available."

Here and there, you see random signs of government officials awakening to the value of the data they're sitting on. In Seattle, Microsoft country, where bureaucrats and hackers can inhabit the same bodies, they've posted a live, self-updating traffic map on the World Wide Web. On the other hand, the lost-and-found database at your local police department isn't quite there yet.

And that walk-in terminal in Arlington? Well, for now it's a working monument to how much money the government can spend to make information hard to get.

DIRTYTRICKS@CAMPAIGN96.ORG

September 1995

Senator Bob Dole may be older and more Republican than the average Internet surfer, but you have to hand it to him—he's right out there on the World Wide Web with a home page for his presidential campaign, at the appropriately official address, **http://www.dole96.org**. Cool! Check it out:

"**This page is here to offer information about Dole's candidacy, his views, and what you can do to help,**" it says. Nice picture of Dole, too. Do we detect a sort of self-deprecating humor in the image of the candidate next to a Dole Fruit Company logo?

Read on:

Bob Dole is often mixed up with the Dole fruit company. In fact, there is no connection between the two, except maybe that Dole (the Senator) is a big fan of tropical fruits, especially slightly over-

> **ripe bananas that are just starting to turn black on the outside, but which are not so black and mushy as to be inedible. Sorry, we're new at this web thing. Anyways, Bob Dole is a staunch Republican. Which is a good thing, what with all those weenies out there challenging him for what is rightfully his—the Republican Nomination.**

Yes, it's already dirty-tricks time in the land of online politics. At least nine Republican candidates or potential candidates have established colorful multimedia billboards in cyberspace—and at least two of these are utter frauds. You have to be fleet of foot, that's for sure. The parodies have stepped out in front of the actual candidates. Soon, the Dole campaign people say, there will be a certified accept-no-substitutes Bob Dole for President home page—otherwise, they haven't quite managed to articulate a digital counterstrategy.

"I wouldn't characterize it as our top communications priority," says Dole's campaign press secretary, Nelson Warfield. After all, we're talking about "a vast wasteland of unregulated Internet communication." (Even in the Republican revolution, *unregulated* can sometimes sound like a curse.)

An ugly little slugfest is brewing here, and like so much in the frantic Wild West atmosphere of the Internet these days, this one mixes the comic, the nasty, and the profound. Just as the notion of online pornography brings sweat to the brows of people who have managed to make their peace with the boobs and goons on television, people well accustomed to traditional political satire are sure to heat up when they see the scary new online version.

•   •   •

The fake Republican Web pages do expose some of the Net's touchiest vulnerable points in the areas of free speech and authentication. So far, it's not hard to tell the phonies from the real thing. Lamar Alexander's is real. Pete Wilson's ("being a glib populist isn't enough in today's political arena") is fake. Phil Gramm's is real—his speeches, his welfare proposal, his "multi-media archive," his automatically updated quote-of-the-day, his red-white-and-blue graphics. At least, I think this is real.

But it's all too easy to imagine the impostors gaining subtlety. The way the Web works—links to new pages constantly being embedded in other people's pages, creating, in fact, a world-wide web of interlinked, cross-referenced data—counterfeits can spread much more efficiently than they can be eradicated.

Then there's the little matter of Internet addresses—the almost-plain-English names that front for the number sequences that identify each computer site. Whereas the gramm96.org address has been acquired by the genuine Phil Gramm for President organization in Washington, dole96.org has been assigned to a "Robert Dole in 96" on K Street in San Diego, and a minor bit of cybersleuthing reveals that this particular campaign juggernaut is actually a twenty-five-year-old network developer by the name of Brooks Talley.

"I'm just an ordinary politically active computer type, doing my best to help out those candidates that deserve it," says Talley cheerfully. "Actually, Gramm beat me to gramm96. I was pissed. To make up for it, I got limbaugh96, buchanan96, and gates96." Talley also turns out to be the evil cybertwin of Pete Wilson of California: "Check out Wilson's new logo. I'm sure he's very proud of it."

Though the satire in these pages is more in the *National Lampoon* than the Jonathan Swift tradition, some people will bite at anything. The fake Dole organization attracts sincere inquiries about the senator's campaign positions, and Talley continues the farce with e-mailed replies like: "We're sorry, but your position on this issue directly contradicts one of our major campaign contributors. Therefore, we have no interest in your suggestion." Yikes. There's a little something for the real Dole to think about when he gets around to handling e-mail.

It would be an understatement to say that the current state of dole96.org points up a lack of regulation, oversight, and standards in the system for assigning domain names. First come, first served has been the basic rule for handing them out to network sites ranging from giant corporations to children's PCs. (Mcdonalds.com belonged originally to the whimsical Joshua Quittner, then of *Newsday*, and there were bloodied noses before the hamburger company managed to gain custody. And for that matter, long before the *New York Times* acquired nytimes.com, one of its more savvy reporters had salted away nyt.com for his Macintosh in San Francisco.)

With demand for domain names multiplying feverishly, the old system has already been overwhelmed, and it remains to be seen whether the Internet can stay free of a slow, costly, lawyer-ridden bureaucracy like the one that protects trademarks. It's a big world: lots of people and companies might have an interest in shoes, but only one can have shoes.com—as it happens, Alan's Shoes of Tucson, Arizona.

At least Alan's really sells shoes. Brooks Talley isn't really running for president (he isn't old enough). The problem of Internet users donning digital disguises has barely begun to come into focus. Maybe you don't mind Talley's having a little fun at

Bob Dole's expense. Certainly no form of speech is more clearly protected by the First Amendment than political satire. But presumably, no matter who you are, you won't be thrilled when people starting posting public messages with *your* signature at the bottom . . .

The very structure of cyberspace changes the rules of how to find people and how to *know* you've found them. In the world that's now ending, there was one (more or less) telephone book listing your one (more or less) true number. In the new world, there is no central directory—or, rather, there are scores, growing and changing daily, along with information-retrieving robots and search engines that can lead you with delightful efficiency to Dole 96. To all the Dole 96s, that is.

For why should the ingenious and fun-loving citizens of the online world stop with one impersonation? There's going to be no Internet moderator ending this game by saying, "Will the real Bob Dole please stand up?" Instead, at least for a while, we will be visiting a hall of mirrors, with many Doles, fat and thin, real and unreal.

MAKING MICROSOFT SAFE FOR CAPITALISM

November 1995

Before he installed Windows 95, John Dodge connected to the Internet using software from a Microsoft competitor, CompuServe's Internet in a Box. Not anymore—Windows 95 silently disabled a key piece of his setup and made it too difficult for him to reinstall it.

Dodge is no novice. He is senior executive editor of the trade journal *PC Week* and so has access to the highest-level support engineers. But life is short and even software professionals learn to take the path of least resistance—in this case, the path leading to Microsoft. He has become a regular user of the new Microsoft Network, though he has trouble with its Internet features.

Still, he believes Microsoft executives when they deny trying to gain market share by sabotaging competitors' software. He just wonders whether Microsoft "has a full appreciation of its actions in the marketplace."

There is reason to believe that Microsoft does.

•   •   •

The government's lawyers are engaged in the third major phase of an investigation that may prove to be the most important, and the most difficult, in the century-old history of antitrust law. Its target is a scrappy, young, fast-moving company with a mere 18,000 employees—a fraction of the size of IBM and AT&T, the last great subjects of antitrust action. Microsoft does not control a manufacturing industry (as IBM did), a natural resource (as Standard Oil did) or a regulated public utility (as AT&T did). Microsoft's strategic monopolies—for it does possess and covet monopolies, despite vehement denials from its lawyers—are in a peculiarly subtle and abstract commodity: the standards and architectures that control the design of modern software.

In a historical eye blink, as the technologies of computing have come to pervade the world's economic life, Microsoft has turned twenty years old. When Ronald Reagan became president, Bill Gates's new company was an unincorporated partnership with accounts kept in handwritten ledgers. Apple was a big new personal-computer company, worth $3 billion; IBM, the mainframe giant, was cobbling together its first personal computer out of parts from outside suppliers. By 1990, just a decade later, Microsoft had become the world's richest software company, though it had no leading product in any important category but operating systems. Today nearly half of the world's total PC software revenue goes directly to Microsoft.

"I personally believe that Microsoft is the most powerful economic force in the United States in the second half of the twentieth century," says Eric Schmidt, chief technology officer of Sun Microsystems—a minicomputer and networking company whose business used to be remote from Microsoft's but now finds itself under direct competitive pressure. Some of Microsoft's control over computing, at all levels, is obvious. Much, however, is invisible. Even longtime insiders are just beginning to understand the nature of that power: how Microsoft acquired it, preserves it, and exercises it.

"The question of what to do about Microsoft is going to be a central public policy issue for the next twenty years," says Mitchell Kapor, the founder and former CEO of Lotus Development Corporation—once the leading PC software company. "Policy makers don't understand the real character of Microsoft yet—the sheer will-to-power that Microsoft has."

The vast majority of the world's personal computers— estimates range from 80 percent to more than 90 percent—run on Microsoft software from the instant they are turned on. Yet, pervasive as PCs are now, Microsoft has made clear that they are only the beginning. The company is working toward wallet computers that carry digital signatures, money and theater or airplane tickets; toward new generations of fax machines, telephones with screens, and car navigation systems; toward Microsoft-run interactive television boxes, office networks, and wireless networks, and, most potently, toward an aggressive Microsoft role in the Internet itself.

By making connections among all these levels of modern computing, and by exerting control over the architectures that govern those connections, Microsoft is in the process of transforming the very structure of the world's computer businesses. "Microsoft is imposing a new verticality on the industry," says Gary Reback, a Silicon Valley technology lawyer who represented a group of anonymous Microsoft rivals in the antitrust proceedings. "Bill's been able to exploit the market far better than anybody else has, and I think that's because he intuitively understands what enormous power he has and how to exploit that power."

It is a software company with the broadest possible understanding of *software*: not just computer code but books, news services, music, movies, paintings, maps, and directories of people and businesses. It believes that you will buy all these online, and it intends to deliver them. With its new Microsoft Network, providing both an online service and Internet access, it is focusing on electronic financial-transaction processing—which is to say, all electronic commerce; which is to say, at least in some visions of the future, pretty much *all commerce*. "Basically what Microsoft is trying to do is tax every bit transition in the whole world," says a senior executive of a competing software company. "When a bit flips, they will charge you."

Its profit margins are staggering by the standards of manufacturing companies—it salts away about a quarter of every dollar that comes in, compared with about 3 cents for Apple. It sits on an enormous reserve of cash. Among modern corporations it has been an unparalleled generator of personal wealth. Never mind that its founder and chairman may on any given day be the world's richest person; the third-richest Microsoft executive, Steve Ballmer, owns close to $3 billion in Microsoft stock, and 2,000 or more of its employees have become quick millionaires, creating a remarkable new class structure in Seattle's social and political life. In a less-charged era, Gates and Ballmer both occasionally joked about their goal of world domination. Now they are more careful. Microsoft's people are taught to avoid using the word *dominate* in public discussion of the company's role in any part of the software business; the preferred word is *lead*.

· · ·

"**T**here are many, many articles that say Microsoft is about to fail," Gates tells me in a hasty interview on the eve of a vacation in China. (He must have Satan's own clipping service.) "Those two extremes are silly beyond belief. We won't fail tomorrow, and we don't have a guaranteed future. That's just logical." It has become an article of faith—with considerable help from Microsoft —that no credible threat exists to its monopoly in operating systems for personal computers or its rising dominance in all PC software. This summer, during the orchestrated buildup for Windows 95, Wall Street found that Microsoft is the company that drives the American financial markets as only IBM and General Motors could in previous eras. The closing months of 1995 see competitors and potential competitors—IBM, Apple, Lotus Development, WordPerfect, Novell—fading back from Microsoft's businesses or bracing fatalistically for the next onslaught.

Gates is back now from his vacation: a personal trip, but he did find time to meet formally with President Jiang Zemin, and Beijing announced—no Antitrust Division there—that it was declaring Windows to be the country's official software standard. Days later, Gates revealed that he had bought the world's greatest storehouse of historical photographs, the Bettmann Archive, adding to his already unchallenged collection of visual images. Two more markets cornered, it seems. The government must ask now, as the computer business is asking, whether a dangerous threshold has been crossed—whether a single force has taken control of the most tempestuous, inventive, unpredictable industry of our time.

CULT OF BILL

By now it's well known that where other companies have offices, Microsoft has a huge and verdant campus—low-slung steel-and-glass buildings set amid stands of evergreen trees in a Seattle suburb, with softball fields and basketball courts and an artificial pond called Lake Bill. Most employees still have private offices, and soft drinks are still free, but the campus has lately taken on an air of relative maturity. It's full; a new set of buildings are on the rise across the freeway. A soccer field was torn up this summer to make way for the extravaganza of the Windows 95 launch.

Everywhere, though, is a sense of the forceful influence of the company's forty-year-old leader, who at One Microsoft Way is always referred to simply as Bill.

Bill's so smart, says a character in Douglas Coupland's new novel, *Microserfs* (another Microsoft first: popular fiction inspired by its wondrous corporate culture). **Bill is wise. Bill is kind. Bill is benevolent. Bill, Be My Friend . . . Please!** A heightened casualness does strain the voices of Microsoft middle executives when they drop mention of "face time" with Bill.

Sometimes people at Microsoft say that they are a mere surfboard and Bill is the man who rides it. The company went through a series of short-lived presidents before finally realizing that a president in the presence of Bill was an impossibility; now there is just an Office of the President, occupied by a group of vice presidents. "One of the things that make us work today is the incredible brain capacity, memory capacity that Bill has," says the most senior of them, Ballmer, striding energetically around a tiny conference room.

Microsoft wears the personality of its leader like a wet suit. Gates's mind-set might be described as a blend of ruthless competitiveness and planned paranoia. He chooses to be scared; he wants his company to be scared. At the moment it is the explosive rise of the Internet that scares him most. At similar critical moments in history—"discontinuities," as he accurately puts it—he has watched most of his competitors stumble and fail, beginning with IBM. He goads his employees with fear of failure. It may help that Microsoft is the company that the rest of the industry loves to hate.

Accusations that Microsoft's people lie, cheat, and steal information are as much a part of the company's lore as its cadre of millionaires with FYIFV ("Fuck You, I'm Fully Vested") buttons. Microsoft knows it has clout, and it uses what it has: to pressure small competitors, trade-show operators, journalists, retailers (shelf space for non-Microsoft software will be at a premium this Christmas), and everyone else.

"Can you name anybody that's happy about being in the same industry with Microsoft?" Mitchell Kapor asks.

Microsoft lives according to a "thin ethics," as he sees it: "Anything not a direct lie or clearly illegal is okay to do and should be done if it advances Microsoft's tribal cause. This licenses the worst sorts of manipulations, lies, tortured self-justification, and so on." Microsoft is hardly alone, of course; plenty of its competitors would play as rough, if they only could. Others in the industry suggest that Microsoft's small-company scrappiness has kept it from facing the issue of corporate ethics: behavior that people will forgive, or at least understand, in a start-up looks considerably less attractive when David grows into Goliath.

Microsoft stumbles, but less often than its competitors; and when its competitors make mistakes, Microsoft has historically managed to take advantage. It has cultivated an aura of inevitability. It has failed so far to overcome some rivals, but it has never lost an important franchise once gained. And if Microsoft people are now openly contemptuous of the government's multiphase investigation of its trade practices, it is Gates who sets the tone. This spring, when the second phase ended with Microsoft's dropping a proposed acquisition of the financial-software maker Intuit, Gates said sarcastically, "In the future we may wait a week or two before we decide to do something like this again."

None of the above appears in the entry on **Gates, William Henry, III (1955–)**, in the world's best-selling multimedia encyclopedia: "Much of Gates' success rests on his ability to translate technical visions into market strategy, and to blend creativity with technical acumen. . . ." There is a picture, too, with a sound clip: "Microsoft was founded based on my vision of a personal computer on every desk and in every home. We've never wavered from that vision."

Needless to say, that's not the *Encyclopaedia Britannica,* now struggling for its life. The leading encyclopedia in the multimedia world is Microsoft's own Encarta—a glossy retread of the old Funk & Wagnalls, updated with pictures and audio bits. Microsoft is rapidly accumulating best-selling entries in every reference category: general desk reference; movie guides; music guides; cooking and wine guides. Most of these were licensed or bought outright, but Microsoft's consumer division is gearing up to produce more and more of its own material for CD-ROMs and online information products. Its new Digital Cartography Lab alone employs fifteen highly trained cartographers and geographers, working on a new generation of digital maps. (Hammond, Rand McNally—are you ready?) Over at the Microsoft Network, a fledgling news staff produces a sort of electronic front page every day.

And by the way, the unabridged version of that famous Gates motto is: "**a computer on every desk and in every home, all running Microsoft software.**"

MICROSOFT VS. THE INTERNET

Not only is the new Microsoft Network software automatically set up for every Windows 95 user; its icons—"MSN" and "The Internet"—are an astonishingly persistent feature of the "desktop" that stares at you from your screen.

"Does anyone know how to get rid of the Internet Explorer icon so that I can put my Netscape Navigator icon in its place??" asks a Windows 95 user on the Microsoft Network. Over on CompuServe, a user says, "I want the MSN icon to go away, but I don't seem to be able to delete it. How do I get rid of the thing?"

That's what Steve Case wants to know, as president of America Online, the most popular commercial online service and one of the companies with the most to lose. "The tens of millions of existing computer owners who are expected to upgrade to Windows 95 won't be offered choices built into their operating system other than MSN," he says. "The operating system for 85 percent of all personal computers is about to become an exclusionary marketing and distribution tool."

He has sent the same message to the Department of Justice. He argues that the operating system is to a computer what the dial tone is to a telephone: the thing you have to use to go anywhere at all. Just as the Antitrust Division eventually prevented AT&T from using its local-telephone monopolies to perpetuate a monopoly in long-distance service, so it should prevent Microsoft from leveraging its operating-system monopoly into the new territory of Internet and online services.

The Internet has forced Microsoft to make a late change in its online strategy. As little as three years ago, when MSN was a vigorously leaked secret, code-named Marvel, the online landscape comprised thousands of hobbyist bulletin boards and just three giant commercial services: America Online, CompuServe, and Prodigy. The Internet, meanwhile, was obscure, academic and seemingly irrelevant to any vision of electronic commerce. Marvel was designed as an America Online–CompuServe–Prodigy killer: a private service that would host proprietary content from newspapers, television networks, Microsoft's own consumer-product sources, and a wide range of businesses with information and products to sell.

Microsoft was not the only company caught by surprise when the Internet burst into public view, and it was one of the quickest to begin a recovery. It took the unusual step of buying a minority stake in an Internet access company and building a nationwide network that customers will be able to use for dialing into the Internet—paying, of course, by the month or by the hour. Gates insists that Microsoft will remain strictly a software company— "We're not in the connectivity business; we're not in the business of owning wires"—but by last year it was clear to Microsoft, as well as the big online services, that Internet access was essential. And Microsoft determined to provide it by means of a single button on the Windows 95 desktop.

But that button is only the beginning of Microsoft's strategy. In a confidential memo to fourteen senior executives last year, Gates described the rise of electronic communication as a "sea change" and warned that in one category, the sharing of documents among groups of coworkers, "embarrassingly we find ourselves somewhat behind one of our old rivals"—Lotus.

It is a revealing document, with a mixture of goading and exhortation, of futuristic vision and rock-hard attention to Microsoft's singular economics. Nothing matters more than persuading users to pay for upgrades to their software. In mature product categories like word processing, he notes accurately, users will not upgrade or switch products merely for the sake of a few extra features, but they will if the new software takes advantage of a sea change. "It takes even more guts," he wrote, "to bet on the Sea Change when you are the market leader but it is the only way to position yourself for massive upgrades."

Every software division at Microsoft is now redesigning its products to take advantage of a world in which every computer can talk to every other. The next version of Microsoft's CD-ROM encyclopedia can be updated live through the connection to MSN or the Internet. For word processors, integration with the Internet means thinking not in terms of personal documents at home or even work-group documents on your private office network, but in terms of browsing, searching, and publishing online. For spreadsheets, it means viewing and manipulating data that comes across private and public networks, interchangeably. "Excel must blow away the competition," Gates urged in the memo. "The basic point, however, is that users' expectation of what Office applications will do is changing and three to four years from now anyone forced to use the software we have today would find it completely inadequate for dealing with the electronic world."

Nathan Myhrvold, one of Microsoft's chief strategists, sums up the attitude now driving every company division: "The Internet is an example of a revolutionary shift that, if we forgot about it, would eventually kill us. The notion that you would do a task on the desktop with desktop software in a few years that didn't involve the Internet is just ludicrous."

Microsoft has already tightly integrated its Internet access into the new Windows 95 environment. Addresses for all kinds of Internet resources can be dragged onto the desktop, where they appear as colorful icons of their own; dragged again into e-mail messages to be shared with friends; and clicked on to begin an automatic dialing process. The Microsoft Network as an online service has its problems—performance is sluggish and the content thin—but as new computers stream into the marketplace with Windows 95 already installed, millions of newcomers will find their way to the Internet by clicking that Microsoft icon.

Hence the extra annoyance of its competitors over the little matter of Windows 95's disabling their users' existing Internet access. Many users who had installed the popular Netscape browser and then tried Microsoft's Internet Explorer discovered that Netscape would no longer work. The same problem affected users' CompuServe's Internet in a Box software.

"Windows 95 includes a process that disables your Internet account," says David Pool, a top CompuServe executive. "And that's just the tip of the iceberg of the inappropriate things Microsoft does from a networking standpoint. It's a clear extrapolation of their operating system monopoly into the network application market."

Microsoft is characteristically unrepentant. "This guy makes me laugh," says Brad Silverberg, head of the personal operating systems division. In the Microsoft version of events, Windows 95 does not "disable" anything. It just happens that some companies' applications cease functioning—they "use nonstandard components" and "need special configuration." Those companies violated Microsoft's published guidelines, he says; they have realized their error and are preparing new versions of the software to repair the problem.

The truth is not quite so innocent. Most Internet dial-up software written for Windows relies on a piece of software called winsock. Everyone's winsock is supposed to be more or less interchangeable with everyone else's, but differences do exist. Many vendors put their winsock into the Windows directory of the user's computer—a friendly practice, since it is then available to other software that might need it, but a risky one, too. If Windows 95 sees a non-Microsoft winsock, it carefully and explicitly replaces it.

"It's not like we blow it away and it's gone forever," Silverberg says, beaming with sincerity. "I think we do a very honest and responsible thing. It's admirable, really."

He acknowledges that the specifications for using the operating system's new dialer were slow in coming but says they are now available to all who want them. And for that matter, he asserts, if Microsoft chose to keep such specifications private, to give a competitive advantage to its own software departments, that would be the company's privilege. It does own the operating system, after all.

MORE WINDOWS, BIGGER WINDOWS

It is conventional at Microsoft to say that success comes from making good products. Microsoft devotes extraordinary resources to improving its technologies. It has effectively stressed "usability" and crisp design. It has recently created a hundred-person research laboratory that resembles a leaner and harder-driving version of AT&T's Bell Laboratories and IBM's Thomas J.

Watson Laboratory. But the quality of its products has been incidental to Microsoft's triumphs over its competitors.

Even Windows 95 shows more awkwardness and instability than the personal operating systems that have long been available from Apple, IBM, and Next. It adopts virtues of all those systems, but many users will still struggle with obscure techniques for allocating memory to their old DOS programs, or find that they regularly crash the entire system. "In many ways this is an edifice built of baling wire, chewing gum, and prayer," wrote Stephen Manes in assessing Windows 95 for the *New York Times*.

It is conventional in the industry to say that Microsoft cannot make great products. It has no spark of genius; it does not know how to innovate; it lets bugs live forever; it eradicates all traces of personality from its software. This view, too, misses the point. Microsoft knows that the technologically perfect product is rarely the same as the winning product. Time and again its strategy has been to enter a market fast with an inferior product to establish a foothold, create a standard and grab market share.

Designing the ideal laboratory operating system and competing in the real world are two problems that have little to do with each other. Apple has had the benefit of a closed battlefield; it could design its software for a limited set of hardware that it controlled. That was a huge advantage for developers and, ultimately, a fatal disadvantage in the marketplace. IBM created in OS/2 an operating system clearly superior to Windows 3.1 in most important respects; yet it failed to persuade the hundreds of crucial manufacturers of PC hardware and the thousands of independent software developers to fall in line with compatible products. Windows 95, despite its "32-bit" fanfare, contains so much vestigial 16-bit code that it makes Intel's new Pentium Pro processor seem to perform badly. But that ugly old code means that users who make the switch will not have to throw out their old software too quickly. Microsoft's genius has been in navigating—and controlling—the fantastically complex ecology of the computer business.

Microsoft's launch of Windows 95 in August, kicking off a planned $150 million marketing blitz, will live in history as a pinnacle of public-relations showmanship in a public relations–driven year. When thousands of onlookers and journalists gathered under the big top on the Microsoft campus or watched nearby on giant screens, the subliminal message was, We can buy anything: Jay Leno (emcee and vaudeville partner for Gates), the *Times* of London (an entire day's run of a once-great newspaper), the Empire State Building (colored lights usually reserved for national holidays). The press made fun, but it was taken in, too, giving weeks of extensive coverage to what amounted in essence to a product introduction—and an upgrade, at that.

Three months later, Windows 95 boxes are stacked high on store shelves, and Microsoft refuses to release sales figures. Anecdotally, it is clear that millions of high-end users have bought the upgrade but that millions of corporate customers have chosen to delay the inevitable headache, particularly when most existing hardware lacks the speed and memory to run it well. It doesn't matter. In the long run virtually every desktop computer will run Windows 95 and its successors. New computers shipping now have Windows 95 preinstalled by default. Applications developers have either stopped developing for DOS and Windows 3.1 or soon will.

Windows has long since stretched the definition of operating system past the breaking point. The original DOS was little more than a thin (and clumsy) layer of hooks that applications could use for reading and writing data to memory, screen, and disks. Windows 95 not only provides a rich environment for controlling many programs at once; it also offers, built in, a word processor, communications software, a fax program, an assortment of games, screen savers, a telephone dialer, a paint program, backup software, and a host of other housekeeping utilities and, of course, Internet software. By historical standards, you get a remarkable bargain.

Some companies used to live by selling such things. Every time Microsoft adds a new feature to the operating system, ripples flow through the software business. When it added a built-in backup program, it instantly destroyed what had been a modest, competitive market in backup utilities; the only customers left were those with highly specialized backup requirements. And when Microsoft asks to license your technology, you may not always find it easy to say no.

One company that tried was Stac Electronics, which had developed software that used a compression technology to effectively expand the capacity of users' disks. Microsoft wanted to build Stac's technology into the operating system and negotiated in its usual scorched-earth style, demanding a worldwide license for a one-time flat payment and threatening to move ahead with or without Stac's license. Stac refused, Microsoft acted on its threat, and unlike most small companies that brush up against Microsoft, Stac sued. A jury, finding that Microsoft had stolen Stac's property, awarded $120 million for patent infringement. Microsoft then swallowed its pride and acquired the technology by settling with Stac, buying a 15 percent stake in the company. Stac now exists as a happy Microsoft partner and the disk-compression business is no more. There are pilot fish that manage to swim with sharks, and there are fish that get swallowed.

A new cycle is beginning: with Windows 95 out, new groups of software companies are struggling to rethink their place in the market. Fax software companies are one example; and if Microsoft has its way, Internet software companies may become another. The Netscape Navigator leads the market now, but after all, Microsoft's Internet Explorer is almost as good, and it's free.

So the operating system has become, from the consumer's point of view, a useful package of software. From a different point of view, however—the point of view of the essential underlying structure of modern computing—the operating system Microsoft owns has become something else altogether: a collection of standards.

WALK SOFTLY, CARRY A BIG API

The age of mass production could not begin until the world agreed on standards for the dimensions of nuts and bolts. The tire and automobile industries coexist because there are standards for wheel sizes. Standards development acts as a catalyst in economic development; the Internet itself emerged when, from the grass roots, open and free standards were created to allow different types of computer networks to talk with one another. All these standards were set by government or international organizations or by industry consortia. No one must pay a royalty or license fee to manufacture a Class 3 fax machine or a keyboard with keys arranged QWERTY-style.

From the point of view of standards, no form of machinery rivals software for the complexity of its interlocking parts—the number of jigsaw-puzzle interfaces between one element and another. In understanding the two-decade history of Microsoft's increasing control over the computer software industry, nothing matters more than its strategic management of these points of interconnection: the creation, marketing, and then manipulation of standards.

Let's say you are an expert at a small company in the infant field of speech recognition, creating technology to turn the spoken word into stored text. You probably got an invitation from Microsoft during the past year to attend a series of meetings. You and your competitors, under Microsoft's guidance, helped create a standard set of hooks into the operating system, a so-called application program interface, or API. No single company in the field had the clout to produce an API that the others would agree on, so there was danger of conflicting standards. But Microsoft did have the clout.

The result: Microsoft, in cooperation with virtually the entire speech-software industry, will release early next year a "Microsoft Speech Software Development Kit," containing "all the necessary tools." Problem solved. Incidentally, in the course of the meetings, Microsoft received and filed away an enormous body of intelligence on the speech-software state of the art and even the specific product plans of your company. That's a risk you had to take.

"I think it's a good thing," says Bathsheba Malsheen, general manager of technology at Centigram Communications, one of the speech-software companies. "To integrate voice and speech into applications is a costly problem." These standards are open, in the sense that they are publicly available.

But in the long run, who actually owns them? "I guess they really are the property of Microsoft," Malsheen says.

Microsoft has a mail standard, called simply MAPI (mail application program interface). It has a new telephone standard, for letting software interact with telephone equipment: TAPI. It is belatedly but feverishly working on a proprietary online multimedia document-publishing standard code-named Blackbird. Microsoft abhors industry-wide standards-setting: its pattern, with increasing consistency, has been to refuse to cooperate with any standards procedures but its own.

"At one time it may have been, hey, the gang's all here and let's have a consortium blah blah blah," says Ballmer derisively. "You can't have things that thrive and get moved forward aggressively if it takes a consortium."

Money on the Internet will require standards. Visa International and MasterCard International managed to set aside their rivalry long enough this summer to announce that they were creating a joint standard for processing credit-card charges across the Internet. Every major player in electronic commerce needs such a standard; until money can flow across the public network in securely encrypted form, online shopping malls and information services remain more experimental than real. Then, a few weeks ago, the alliance broke apart.

MasterCard, along with Netscape and IBM, charged that the standard, created by Microsoft and published as an "open" set of specifications, was actually proprietary, designed to give Microsoft a powerful advantage, perhaps enabling it to take a slice of every transaction. Microsoft responds that the specifications are freely available; its own Windows implementation of those specifications, however, is proprietary and available for those who wish to pay for a license, possibly on a per-transaction basis. It has become a familiar scenario: Microsoft claims an architecture is public and open; its competitors say the crucial details are reserved to Microsoft alone.

Microsoft is by no means the only company that seeks to exploit private standards. Netscape itself is playing a dangerous game with the standards that gave rise to the World Wide Web:

creating proprietary "extensions" that work only with its own software and hoping that its market dominance will be enough to make them stick. The history of IBM's downfall in the PC industry is a history of failed attempts to impose standards by fiat. IBM took its clout for granted. Microsoft gives top priority to its standards-setting; it "evangelizes" its standards, using every possible form of persuasion to bring the industry in line.

Ultimately, only one kind of company can play the standards game risk-free: a company with a monopoly. The risk for everyone else is that the company that owns the standard can change it without warning, give its own programmers special advantage and freeze innovation elsewhere.

"We've lost this notion of a public standard as good," says Alex Morrow, general manager of architecture and technology at Lotus. "Instead we have this new thing, a quasi-open private standard that's controlled by one company. That's where innovation is going to suffer."

The ultimate standard—the ensemble of standards—is of course the operating system itself: the power spot in the digital ecology. The case against Microsoft, in the eyes of its rivals, comes down to one central issue: leverage, using the operating-system as a fulcrum to gain power in new markets.

The market in big desktop applications is a much-disputed case in point. Not long ago, WordPerfect led the word processor market with a much-loved product and a toll-free customer support service (something Gates has never authorized at Microsoft); Lotus 1-2-3 dominated the spreadsheet market, and Borland International's Paradox led the PC database market. In 1991, Mike Maples, a senior Microsoft executive, described the company's goals in the aggressive style that its top executives used to favor: "If someone thinks we're not after Lotus and after

WordPerfect and after Borland, they're confused. . . . My job is to get a fair share of the software applications market, and to me that's 100 percent."

For all three companies, the fatal "sea change" was the transition from DOS to Windows, particularly Windows 3.0, the first widely popular version. Microsoft notes with considerable justice that its rivals made a strategic blunder in not releasing Windows versions of the software more quickly. Microsoft's applications group and its system group were able to "fly in formation," as Ballmer puts it (zooming his hands cheerfully through the air). Microsoft critics have said that flying in formation included sharing technical information that gave Microsoft's own programmers an advantage over outsiders trying to write fast and well-integrated Windows software, and there is truth to that. But there is also no question that WordPerfect, Lotus, and Borland were late by choice—in part because, caught up in the Catch-22 of the operating-system wars, they knew that their Windows versions would help Microsoft by cementing the establishment of Windows.

The flow of inside information will remain a critical issue for the antitrust investigators. In the eighties Microsoft executives often spoke of a "Chinese wall" between the systems group, responsible for DOS and Windows, and the applications group, responsible for the programs that ran in those operating environments. Ballmer himself once said there was "a very clean separation"—"It's like the separation of church and state." Competitors were dubious, knowing that all neurons at Microsoft led to Bill Gates. These days Microsoft executives take a different tack. They deny that the concept of a Chinese wall ever existed. They admit that their own developers sometimes get an edge in knowing how to take advantage of new Windows features before the knowledge spreads to competitors, but they insist that the knowledge does spread sooner or later—because it is in their interest to make sure that everyone writes for Windows—and they say that's as level as the playing field needs to be.

The final blow to the applications market came with the emergence of "office suites"—packages of word processors, spreadsheets, and databases bundled together. Again, Microsoft saw the opportunity first and made sure that its package was more tightly integrated than its competitors' could be. It announced a new standard, called OLE (for "object linking and embedding"), that allowed, say, a word processor document to display and even work with a spreadsheet. Again competitors charged, and continue to charge, that Microsoft manipulates the OLE specifications to its advantage—changing them to suit its applications programs. Almost as an afterthought, Microsoft also added its not-well-regarded PowerPoint presentation-graphics software to the package, effectively cutting the price to zero and transforming that business overnight. Though *transforming* may not be the perfect word. "Microsoft didn't transform the market, but strangled it," says Karl Wong, director and principal analyst at Dataquest, a research company.

Today, Microsoft says it "leads" the market in office suites. Yes, indeed: its market share is estimated at 90 percent, closer to Mike Maples's target than he could have dreamed four years ago.

FOR ITS OWN GOOD

The essence of antitrust law is an American view that the public has an interest in preventing excessive concentration of economic power. In the sixties, two companies appeared to have such power, in the two industries with the greatest grip on the future, computing and telecommunications. The investigations of those companies, through several presidencies, formed an era in antitrust law that ended abruptly on a single day: January 8, 1982. The Justice Department dropped its long-running case against a jubilant IBM but announced at the same time that AT&T had, with bitter reluctance, agreed to a historic breakup.

Today, IBM has lost sway over every business it participated in. It allowed the PC industry to emerge at its feet, and it turned itself from a paragon of financial reliability into a company that for several years was losing money at a frightening rate. It has become a stagnant noncompetitor, looking for ways—its only hope—to break itself up into smaller business units.

At AT&T, meanwhile, it is now an article of faith that the court-imposed breakup was a brilliant turning point in the company's fortunes. It was the event that freed it from its own hamstrung indolence and enabled it to compete in new arenas. AT&T is continuing what the government began, breaking itself up into smaller and, it hopes, more agile companies.

Monopolies become their own worst enemies—particularly in businesses that live or die by technological innovation. They get soft. They make poor research choices. They bleed both profit and invention. They poison the marketplace that created them. In the rarest cases, like AT&T's, an outside force can save a monopoly from itself, but government interference is always frightening and never popular.

It's certainly unpopular with many politicians—witness the "pinch me" statement, a comment by Senator Bob Dole that Microsoft rushed into its legal briefs and news releases: "Let us understand what is going on here. A company develops a new product, a product consumers want. But now the government steps in and is in effect attempting to dictate the terms on which that product can be marketed and sold. Pinch me, but I thought we were still in America."

Microsoft's lawyers encourage an ideological view of *United States v. Microsoft*, employing not just "free-market capitalism" arguments but also a quaint form of red-baiting, assailing would-be "commissars of software," and insisting: "Such thinking should have disappeared with the Berlin Wall. Fortunately for American consumers, we do not have a centrally planned economy."

"It's like a throwback to the 1950s," says Case at America Online: 'What's good for General Motors is good for America.' "

For her part, Anne K. Bingaman, assistant attorney general in charge of the Antitrust Division, bridles at suggestions that the political climate could affect the investigation—and also at a widespread industry view that, in the end, the high-technology business will prove too fast-moving and too technical for the non-nerd lawyers in Washington to keep up.

"We have a much better handle on the industry than people realize," Bingaman says. "The group of people that work on these matters have long and deep experience. We keep up. We understand it. We have sources."

Bingaman is proud of achieving the consent decree in phase one, in which Microsoft agreed to end a set of licensing practices without admitting any wrongdoing or suffering any penalty. The most blatant was an arrangement in which PC manufacturers paid Microsoft the same royalty for shipping a computer without DOS as with DOS—meaning that, if you were one of the few people who bought a non-Microsoft operating system, you paid its manufacturer and then you paid Microsoft on top of that, a huge disincentive. Microsoft was "locking up the market with practices which every computer manufacturer despised and which the competitors despised," Bingaman said in July 1994. "To get these low prices you had to sell your soul and never leave Microsoft." And she also said: "I hope consumers, within a short period of time, will have more choice of operating systems."

It has not happened. The practices Microsoft agreed to forgo had already served their purpose. Gates was right when he summed up the effect of the consent decree in one word: "Nothing."

And each month brings new issues, all variations on the same theme: Microsoft's use of not-quite-public standards as a sword and a goad. The Microsoft Network was shipped with every copy of Windows 95 before the government's lawyers could decide whether to act. Now they must consider the new Microsoft-Visa agreement on standards for financial-transaction processing— open standards, according to Microsoft; closed standards, according to MasterCard and Netscape—and as of this month Visa has already shipped its Windows software implementing the standards. "We are not giving away our implementations of those specifications, just as we don't give away Windows or any other software product that we make," says Craig Mundie, senior vice president of Microsoft's consumer systems division. Microsoft is well along in the creation of proprietary software to handle every stage of the process, from customer to merchant to bank.

Meanwhile, the stores are filling with third-party software boxes displaying the official Windows 95 logo. To get Microsoft's permission, the manufacturers had to demonstrate not only that their software runs under Windows 95, but also under the more advanced version of the operating system, Windows NT—a version that so far, despite all Microsoft's evangelizing, does not have the support of many popular applications. That logo is a powerful lever, applying power from one product line to another, and it deserves the Justice Department's attention.

So does Microsoft's new campaign on behalf of not-yet-available online development tools, like the one code-named Blackbird, for companies that want to publish news, design games, build shopping malls, or deliver entertainment over the Internet. Designers of competing tool sets—Netscape and Sun—see Microsoft's as attempts to gain control of another key choke point in the pathways of electronic commerce.

So does an odd bit of language in Microsoft's contracts with the computer makers who bundle Windows 95 with their hardware: a forced promise not to sue Microsoft or anyone else for patent infringement. It happens that Microsoft is building up a strategic portfolio of software patents, both home-grown and licensed.

And so, of course, does the intimate connection between Windows 95 and Microsoft-brand Internet access: the bundling of the Microsoft Network software; the persistence of the desktop buttons; paradoxically, all the features that make online access easiest for new customers. As Microsoft vehemently points out, every other big online service manages to gets its software into your mailbox and bundled with your new computer. Still, the Antitrust Division should, if nothing else, see a natural analogy with the bias AT&T created for itself before the days of equal access. Customers could use MCI—but only by dialing a slightly inconvenient code.

Microsoft retorts angrily to all the telephone analogies by noting that AT&T was a government-regulated monopoly. The folks at One Microsoft Way are merely . . . successful. They are big. If they are linking together pieces of the hardware-software-network chain, they are doing it in a way that has lowered prices, added value and made life easier for consumers. It is not their fault if the economics of scale in the software business, combined with tactics that press antitrust law to its limits, brings them huge benefits.

Is Windows an open standard? Yes—when and only when that suits Microsoft. "We could say, hey, we're not publishing any APIs to our operating system," Ballmer says. "Or we could pick five guys and tell them what's in this operating system—we're not going to tell other people."

And that is where the government should draw its line.

There was a moment in history, just a few years ago, when any number of operating systems, real and imagined, could have emerged to run the world's personal computers. That moment is past. The Microsoft architectures have established themselves so deeply in every segment of the computer business that they cannot be displaced, not even by Microsoft. Those standards are an essential facility—to use antitrust jargon—like the 60-hertz AC current that flows to every American household. To date they have remained mostly open and mostly public, because that served Microsoft's business interest. Now the government could, and should, declare a public interest in open standards in computing.

The Department of Justice does not need to break Microsoft apart. It need only—a far-reaching step in itself—require Microsoft to make its operating system, and the web of standards surrounding it, truly and permanently open. Other companies should be allowed to clone it if they could; Microsoft should be restricted from taking internal advantage of new changes until they were published to the rest of the market.

For that matter, Microsoft should open its standards voluntarily. It will not, but it should: end the painful cognitive dissonance that comes from proselytizing for open standards and then threatening to close them at will.

"It's not like everyone and their brother is going to go out there and beat them," says Eric Schmidt at Sun. "They'd probably have 95 percent of the market anyway. Then all the arguments about their behavior would stop. If they really did open interfaces, it would change the dynamics of the industry in a positive way." It would be for their own good, he says: "They could get back to work and try to build great products and compete."

Gates is right about one thing: Microsoft's future is no more assured than was IBM's. The Internet does pose a threat—a new set of open standards that, so far, Microsoft cannot control. And Microsoft's own power poses a threat, too—the threat that comes with the self-fulfilling destiny of any monopolist. Microsoft could fail to drive consumers to new waves of upgrades; it could stagnate financially even as it retains its grip on the neck of the market. "The company in some sense is a captive of its own history of voraciousness," as a former Microsoft executive says. It is a captive of shareholders who have come to expect nothing less than Microsoft-style profit margins and growth rates. It is a captive, to its own horror, of lowest-common-denominator design, the inevitable consequence of serving a market of 100 million.

The rest of the industry is captive, too. No company has Microsoft's power to place bets; few companies can afford to chance a new approach in a product category near the ever-advancing boundary of Microsoft's Windows package. No quantity can be harder to perceive and harder to measure than innovation that never occurs—the absent pioneers, the fading of vitality in a still-comfortable industry.

No monopolist wants to be relieved of its burden. To

Microsoft, it would be nothing short of theft. They own that operating system—they sweated, invested, and fought for it. If they can put a computer on every desk and in every home, all running Microsoft software—and all connecting to the Internet—consumers should be grateful.

"You click a button and it's so easy!" Silverberg says, grinning again. "How could there be anything wrong with that?"

REALLY REMOTE CONTROL

December 1995

There is a hot tub at 305 Maple Street in Ypsilanti, Michigan, and perhaps you'd like a status report. The water is "nice and warm," 101 degrees, the backup battery is delivering a safe 9.4 volts, and the cover is closed. Over in the refrigerator, it's reportedly 35 degrees and dark, and the last can of soda is gone, but let's stick with the hot tub for a second, shall we?

Drifting along the Information Superhighway, we have reached a site belonging to a man named Paul Haas. Web pages are at best imperfect mirrors of their creators, but it seems safe to surmise that Paul cares deeply for his hot tub. He has run a fifty-foot cable to it from his computer. When someone like me, at home with my own computer, stumbles into his node on the Internet, a signal runs down the cable, activating a set of sensors in the hot tub, and, as data streams back, Paul's computer is kind enough to translate it into English for the world to see.

In the name of thoroughness, Paul has also provided a set of Frequently Asked Questions about his hot tub, which is an octagonal whirlpool spa big enough for eight soakers, and for some reason my eye jumps right to No. 6: **What Is the Point?**

Would you believe . . . safety? "I can check the hot tub from inside my house or at the office," he explains. "It gets cold in the winters here. If the heater fails, the water could freeze and damage the tub." Sure, Paul. And I suppose it's just ordinary prudence that led you to set up that robotic hand, attached to a servomotor, wired back to your computer, so that someone as far away as Oslo or Kyoto can select the micro, royal, normal, or tidal strength and wave to your cats?

No—what's happening here is stranger than that. Not only is the Internet turning us all into one big, noisy cocktail party; it seems also to be assembling a big, gangly machine, with robot arms and video eyes all over the world. There are microphones and CD players, pointable telescopes, and at least one paper shredder. So look around. Push those buttons.

This is a case of *something* expanding to fill the available space, just the way automobile traffic expands to fill new highway capacity and software expands to fill your hard disk, no matter how big your hard disk gets. In this case, the available space is the world's bandwidth—the capacity of the wires joining our homes, offices, and computers. Bandwidth used to be expensive. Telephone companies behave as though it is still expensive, and it is when you make traditional voice calls across transoceanic fiberoptic cables or satellite networks. But the economics of the Internet are different, subversively so, and here you are using that same fiber for essentially no charge. It's a good thing, too, because it would be embarrassing to spend too much money pressing the **Go** button that has been drawn on your computer screen by the Interactive Model Railway group at the University of Ulm, Germany.

Maybe you choose the Märklin E 94 on platform 1; your command passes to a tabletop train controller over in Ulm. The train starts to chug. A live video camera sends a picture back to your screen, along with a schematic diagram. Playing trains may not have been what AT&T had in mind when it laid that cable under the Atlantic, but the bandwidth is there and people are finding ways to use it.

The attaching of devices to the Internet seems to have begun with college soda machines, generally in the vicinity of computer-science departments. It may have been useful to check contents and temperature without having to stir from your keyboard fifty yards away. Presumably the utility declines rapidly with distance, but that has not stopped some of these machines from achieving a kind of celebrity status—holy resorts for online pilgrims.

Along with the ingenuity, it's possible to detect a serious form of frustration at work here. Thousands of people weekly use an Internet Pizza Server ("**Good News!—The Internet Pizza Server has finally been accessed from every continent on Earth! Thanks to Chris Jung at McMurdo Sound, Antarctica, for being coerced!**"). They choose size and toppings. They can get half-and-half pies. They wend their way through an elaborate tree of menus and project descriptions, and by the time a heavily engineered software engine delivers the final product, you begin to suspect that they've actually forgotten that there is a difference between a pizza and a *picture* of a pizza.

Those who stumble across the Automatic Talking Machine at Inference Corporation can have a sentence or two synthesized and spoken aloud to one Rob Hansen, owner of the device. Or to his empty office—you never know. Lately, online wanderers have used the ATM to say:

```
Hi there! I wonder if anyone can hear me
over there in Los Angeles. If you can,
then I would like you to know that it is
a little bit cold here in Newcastle, Aus-
tralia, tonight.

Hello.

This has been and still is Oxford. It is
full of buses and pollution. There also
seems to be a distinct lack of reality.

Mr. Robert Hansen, this is the police, we
have the building surrounded, come out
with your hands up.
```

Let it not be said that the evolution of the Internet as communications tool is yet complete.

Is this power or is it powerlessness? Maybe it's power if you were looking for a quick way to page Bindu Wavell on campus in Colorado, or eavesdrop on Erik Nygren's CD player at MIT, or water and watch the seedlings at the "Tele-Garden" in Irvine, California. But you weren't. It's a reminder of how embryonic and open-ended the online world remains.

Still, you can check current stream-flow conditions in California, courtesy of the United States Geological Survey, or beam positions and vacuum currents at the Cornell High Energy Synchrotron Source. You can peer through scores of spy cameras—probably the fastest growing category of plugged-in devices.

George Orwell, of course, provided us with a sour advance guess at what a world of internetworked television cameras might mean. So far, it seems to be turning out differently. Some people seem to *want* to be watched—and even talked to—by the random online stroller. Even more seem to want their pets to be on parade: you can while away the hours watching someone's iguana or ant farm. Mostly, though, the cameras are outdoors, and the possibilities run from live sunsets in Bozeman, Montana, to views of the main Berlin railroad station, to innumerable shots of highway traffic, to the scene at the 18th hole of the Silversword Golf Course in Maui.

You can not only watch but manipulate the arm of a six-axis robot at the University of Western Australia. The robot's masters report that "thousands of users have demolished block towers and a smaller number have stacked blocks." *Much* smaller is my guess, based on a few minutes of trying. When you're done, you may or may not feel an odd tingling in your arm. Either way, it seems prudent to skip past that last command: **Nuke mainland France.**

HERE COMES THE SPIDER

March 1996

Over in the real world, robots are not doing very well, at least by the standards of 1950s science fiction, where by now they were supposed to be doing our dishes and vacuuming our floors. But in the virtual, electronic, online world, robots are thriving. In fact, as soon as there was a World Wide Web, it was inevitable that there would be Spiders. They crawl about the Internet, looking for information wherever they can find it. They work by night or by day; they leave small telltale traces of their visits; occasionally the bad ones *go berserk* and *wreak havoc.*

Sending virtual robots out to make sense of the Internet is a natural idea in the same family as sending industrial robots to clean up nuclear reactors—scary, messy jobs likely to be hazardous to human health. The essential problem is one that never existed before in library science or information management. People responsible for the contents of the Library of Congress or the archives of the *New York Times* or the centrally managed services of America Online tend to know what they have—or at least where to look for what they have. Their information has boundaries and limits. It resides in a certain place. On the Internet, information is limitless and scattershot; it resides anywhere and everywhere; and no one is responsible.

You are, let's say, Karl-Eric Tallmo, a writer and musician living in Stockholm. You create a site on the Web. Maybe the computer you are using for storage is online most of the time. You include—publish—things you like, by friends and strangers. Someone who stumbles across your site can find poems (in Swedish) by Linda Hedendahl and paintings by Kent Wahlbeck. You enjoyed the recent television production of *Pride and Prejudice,* so . . . why not? "Here's the original novel written in 1813—all three volumes!"

You are probably not the first person to attach Jane Austen's text to the Web, but how would we find out? We would use one of several Internet search engines, all more or less experimental. Here goes: we search for **truth universally acknowledged** and **Darcy**. We find more than 4,000 matching documents, including these:

- **"It is a truth universally acknowledged, that a single man in possession of a good fortune, must be in want of a wife** . . ." followed by the remaining complete and unabridged 676 kilobytes of *Pride and Prejudice,* stored on the hard drive of a computer in Japan.
- Excerpts, provided by an electronic-book company attempting to sell Austen's entire text (among other "Romance" titles) for a fee of $2.25.
- More than a hundred answers, accumulated at a site in Alberta, Canada, to the question, "Why Did the Chicken

Cross the Road?" Answer No. 3 is: "**Jane Austen: Because it is a truth universally acknowledged that a single chicken, being possessed of a good fortune and presented with a good road, must be desirous of crossing.**"

- The "statement of purpose" of the Princeton Pacific Asia Review: "**The strategic importance of the Asia Pacific is a truth universally acknowledged...**"
- An article about barbecue, courtesy of the Vegetarian Society UK: "**It is a truth universally acknowledged among meat-eaters that...**"
- The home page of Kevin Darcy, Ireland. The home page of Darcy (**Cool 'n' Dangerous**) Cremer, Wisconsin. The home page and boating pictures of Darcy Morse. The vital statistics of Tim Darcy, Australian footballer. The résumé of Darcy Hughes, a fourteen-year-old yard worker and baby-sitter in British Columbia ("**preferred location of employment: North America**").

And it's a moving target: now the same search request will turn up this very text, filled as it is with Darcys and universally acknowledged truths.

Each time we enter a search request, we naturally cannot be searching the entire universe of interconnected computers. We are searching a previously compiled index, and it is robots who do the compiling. The Web now comprises hundreds of thousands of sites, with tens of millions of individual "pages" and tens of billions of words—best estimates produced, of course, by robots. No doubt the visitor's log of Karl-Eric Tallmo's computer, like the logs of mine and just about every other site on the Web, has lately been accumulating notations from interlopers called "Scooter" and "Spidey." They read everything they can find.

"Scooter" is the creation of Louis Monier, a Paris-educated software engineer at Digital Equipment Corporation's laboratory in Palo Alto, California. Scooter is not yet in regular operation, but it has run around the Web a few times, most comprehensively in December and again in February, compiling an index that Digital makes public as what seems to be the most comprehensive Internet search tool available, AltaVista. Scooter does roughly what you would do if you could set up a thousand PCs, all running Web browsers, and hit your mouse about thirty times a second, taking notes furiously all the while.

Robots need to be capable of some decision making. They construct a map of the Web's topology. It is important not to keep visiting the same Web pages again and again—not only does this waste the robot's time, but it ties up bandwidth and irritates owners of Web pages. These days a well-behaved robot looks for a standard message that people can use if they want to warn robots away from their sites. But robots and spiders don't really move— not even electronically. They have not begun to approach the visionary conception of intelligent agents: software assistants that might truly wander from computer to computer on your behalf. "This notion of a program that leaves your machine, runs on different machines and tries to find you the cheapest pair of sneakers," says Monier, "there's a name for that—it's a virus."

Even so, perhaps it's not too much of a stretch to pretend that these software projects are creatures of a sort. In their rooting about, they are assembling something new: a thing that, with all its randomness and idiosyncrasy, all its dead ends and stranded links, comes closer than any catalogue or reference book to a true guide to our collective human knowledge. "We have a lexicon of the current language of the world," says Allan Jennings, project manager for the Digital search engine.

An odd bit of text that you could stumble across on Karl-Eric Tallmo's Web site puts it almost the same way: "There is no practical obstacle whatever now to the creation of an efficient index to all human knowledge, ideas and achievements, to the creation, that is, of a complete planetary memory for all mankind."

Those are H. G. Wells's words, written in 1937. "It foreshadows a real intellectual unification of our race," he continued. "The whole human memory can be, and probably in a short time will be, made accessible to every individual. And what is also of very great importance in this uncertain world where destruction becomes continually more frequent and unpredictable, is this, that . . . it need not be concentrated in any one single place. It need not be vulnerable as a human head or a human heart is vulnerable. It can be reproduced exactly and fully, in Peru, China, Iceland, Central Africa, or wherever else. . . . It can have at once, the concentration of a craniate animal and the diffused vitality of an amoeba." Wells had no inkling of computer networking, of course. The hot new technology that inspired him was microfilm.

MANUAL LABOR

April 1996

Good morning—Daylight Savings Time has begun, and you have work to do. Maybe you have already reset your wristwatch and bedside clock, but the odds are that you have neglected one or more of the following time-keeping devices: VCR. Automatic coffeemaker. Telephone answering machine. Camera. Digital thermostat. Television set. Microwave oven. Stereo. Personal computer. Car.

Of course, these appliances have other functions, but they generally contain clocks, and, as a citizen of a technological age, you may feel a responsibility to keep your technologies in order. Or you may not—your household may contain more than a few LCDs permanently lit with that flashing **12:00:00**. Either way, you're well aware that the days when you could set a clock by pushing the minute hand with your forefinger or turning a knob are long gone. In fact, for some of these machines, you are going to have to find—and at this point a shudder is understandable—the instructions.

Few tasks so perfectly crystallize the problem of inhuman design—design for engineers instead of for users—as the simple business of setting a clock. There are rental cars all across America that will never see the correct time on their dashboard clock radios, with their tiny unlabeled buttons and knobs. There are answering machines that tag phone messages with the correct time from October to April only; from April to October, their owners resignedly live with the one-hour error.

Too many functions—not enough buttons. They get over-loaded. The result, a form of pidgin devised exclusively for instructions:

```
When the Preset/User selector is set to
Preset, this button acts as the recall
button which recalls the factory default
setting. When the Preset/User selector is
set to User, this button acts as the pro-
gram button. When pressed, the user may
program the present display mode as any
one of the unused User modes, or overwrite
a previously programmed User mode.
```

Or too many buttons: a typical VCR can no longer be controlled *without* the remote control, and the remote control has forty-one buttons, two rocker switches, four cursor-control keys, a "TV-VTR" toggle, and a big, complicated "REW-FF" adjustable-speed knob. Even so, none of those controls has anything to do with setting the clock. That function is relegated to a computer-style menu that the user navigates by scrolling and "executing." The industry, all too aware of those flashing 12:00:00s, has fought back lately with new features meant to automate the whole process, in some cases using a time signal broadcast by public television stations. Now, my instruction manual (have you found yours yet?) devotes two of its 64 pages—14 numbered instructions, 12 illustrations, and 4 bonus small-type Notes—to explaining the use of the "Auto Clock Set feature." Be patient. Be very patient. A helpful Note instructs:

If nothing happens even after you wait about 30 minutes, set the clock manually.

The craft and business of technical writing have bloomed in recent years. The world's novelists are vastly outnumbered by writers whose mission is, pure and simple, *how to.* There are fast-growing university programs, academic journals, consulting firms, and professional societies—the Society for Technical Communication alone has more than 20,000 members. Still, as you wander from appliance to appliance, manuals in hand, you may just occasionally feel that there is room for further progress. You may sometimes sense that, while each individual word is English, you are gaining fresh insight into the syntax and tonality of a foreign, probably Asian, language. You may marvel at the mingling of the obvious and the obscure . . .

Here's how to pick up your handheld vacuum:

Remove the DUSTBUSTER from the charging bracket by gripping middle of handle and gently pulling unit from bracket (Figure 6).

What to do with it when you're finished vacuuming:

> **The DUSTBUSTER will automatically turn off when your thumb is removed from the button.**

Here is the dryer. Start with the "interior light":

> **The light inside the dryer drum comes on automatically when the door is opened. When the door is closed, the light turns off.**

Easy! But they're just softening you up for the body blow of the "wrinkle-rid/cycle signal":

> **This knob controls the volume of the audible cycle signal for both the wrinkle-rid feature and the regular end of cycle signal . . . It should always be turned to an on position when using the wrinkle-rid feature which is built into the Knits/Perm Press Auto Cycle.**

At least it has no clock. And this is old technology. New technology means silicon, and silicon seems to engender instructions along the lines of this, for a cellular phone:

> **There are three ways to select a memory location. You can select a specific loca-**

tion number; you can choose to autoload which will store the information in the next sequentially available location; or you can arrange the location into blocks.

Authors of manuals also find that, like it or not, they have company attorneys as writing partners. This is why a typical manual may contain three black-bordered type boxes headed "Caution: Product Damage," four "Warning: Electrical Shock Hazard," two "Warning: Personal Injury Hazard," and one "Caution: Floor Damage," each with an oversized exclamation mark inside a bold triangle. It is an American phenomenon, mostly.

"European product-liability laws don't make it necessary to warn against abnormal use, like using a TV set to melt the snow in front of your house," says Peter Rich, a Danish consultant. He says lawyer-driven manuals can in fact cut down on lawsuits: "Nobody can find out how to use the product. It is consequently put aside and causes no harm."

More often the instruction manual itself is put aside and causes no harm. It is an article of faith in the consumer electronics business that, no matter how much companies invest in manuals, their readership remains . . . infinitesimal. "We think it's pretty much nobody," says Bill Cubellis, marketing manager for Sony home video products, "but once in a while people will browse through." It is the same in the software business, responsible for an equally numbing sort of linguistic fog. Before there was computer software, who could have predicted that a sentence like the following—recent winner of a worst-of-the-month contest run by Corecomm, a Houston-based technical-writing company—could ever arise in the universe of human discourse?

Type the field name Name in the Field Name field.

Words of one syllable, too. You find yourself staring at them, or through them, the way you would stare through one of those 3D stereograms, hoping that some image will come into focus, if only for an instant. It's almost poetry.

HALL OF MIRRORS

May 1996

There's absolutely no reason to be confused about advertising on the Internet. It's simple. Let's say you happen to be looking at a Web page belonging to a software company like Microsoft. You might see a promotional link to Yahoo, a service that indexes thousands of Internet sites. You click on it.

As the Yahoo page arrives on your screen, you can't miss a discreet ad for, today, Southwest Airlines. Hurry on, though. Let's say you burrow through the Yahoo site looking for "computers." Now a banner ad appears for a service called C|net. So you click on that. And onward you go: a banner leads to Netscape; there, a plug leads back to Yahoo (bought and paid for, though there's no way for you to tell); a new banner leads to Toshiba, then to RealAudio, then back to Microsoft. Loops within loops!

It's hall-of-mirrors time again—circular and incestuous. As pervasive as marketing is in the real world, at least you can usually tell the advertisers and the publishers apart. Not on the Internet.

"A publisher is anyone who commands people's attention,"

says John Houston, chief technology officer of Modem Media Advertising in Westport, Connecticut. "It creates a strange interdependency—what do they call that in biology, a food web? It creates a strange media food web."

The whole business of advertising is being reborn on the Internet—every week, it seems—amid a frenzy of new styles and technologies. Hundreds or thousands of start-up companies are trying to grasp territory, with ever more ingenious services and software. At the moment the Internet commands just $50 million of the $133 *billion* believed to be spent yearly on advertising; there is room for growth.

The most successful current "publisher"—the Internet site pulling in the greatest revenue from advertising—appears to be not a publisher at all, but a software company, Netscape. The company reported earnings of $2.4 million in the last quarter of last year, and according to Webtrack Information Services, publisher of *InterAd Monthly*, about three-quarters of that amount came from the sale of advertising on its Web site. This year's numbers will be much higher: Yahoo alone is paying $5 million for an ad that doesn't look like an ad.

Yahoo, in turn, lives entirely on advertising revenue. For Internet old-timers, the lesson is: There goes the neighborhood. Is cyberspace really descending to the level of any old baseball stadium—or worse, television?

It may just be inescapable. Successful Internet companies receive millions of electronic visits each day. Yahoo's visitors are mostly trying to find their way around the Internet. Some of Netscape's visitors are trying to download free versions of its Web-browsing software; and many are just using that software—by default, it connects to the Netscape site each time it starts up.

The publishers log all those connections, or "hits," or "visits," or "page views," or "impressions"—the language is as variable as the counting techniques. Prices vary, too, but by and large, each time you glimpse a little banner atop someone's Web page, the advertiser has just kicked in another 20 or 30 or 80 cents. That is considerably more than advertisers pay in traditional media— the Outernet, as some have begun calling it. But then, advertisers on the Internet get a valuable kind of instant feedback—known as "clickthrough"—from their mouse-wielding targets.

An amazing proportion of Internet advertising goes to the few leading sites: last year, the top five publishers got half of the revenue, by *Webtrack*'s estimates. That is changing fast, however. Everyone on the Web is a publisher. Everyone, with or without commercial intent, is vying for the same precious neuroseconds that make up the collective Internet attention span. If you have a Web page that draws fifty visits a day, then you, too, can sell advertising.

So you have to get those visitors. Short of paying Netscape $5 million for promotion, what can you do?

In ancient times, it was enough to post photographs of your pet gecko. Now that the novelty has worn off, some Web sites are thriving by inventing services—catering, for example, to people who want to publicize their Web sites. Scott Banister started Submit It, a free, automated resource for bringing your page to the attention of many Web-searching outfits at once.

"A lot of people mail me—they're depressed because they don't show up until the third page of the search results," Banister says. "They're tremendously worried about this kind of thing." So some people cheat. Some of the robots that compile Web indexes can be fooled by brute force: "If you're selling surfboards," Banister says, "you put the word 'surfboard' down at the bottom of your page five hundred times."

Like so many Web innovators, Banister began his service because he needed it himself and because "it seemed like a cool thing to do." Now, of course, he's selling advertising.

The best rates go to publishers who can deliver specialized audiences—groups known to be interested in, say, computers or, even better, networking software. For a price, the Web-search services will actually let you sponsor a word. If you search for *golf* at Yahoo, an ad for iGolf offers to let you win a set of clubs. Meanwhile, over at Yahoo's chief competitor, Lycos, "golf" is owned at the moment by Cobra Golf Incorporated: click here and win a driver. AT&T and Sprint have both bought *telephone* here and there. Search just about any site for *Netscape* and you'll see a Netscape ad.

"There are premium words and less premium words," says Karen Edwards, marketing director for Yahoo. She says that her company has adopted a policy of selling company names only to the company itself, not to competitors. Other services are not so finical. Search for *Microsoft* or *windows* at Lycos, and you get an IBM ad. And IBM is evidently leaving no stone unturned: it has also bought the word *Gates.*

THE END OF CASH

June 1996

Cash is dirty—the New Jersey Turnpike tried to punish toll collectors recently for wearing latex gloves (thus giving the driving clientele a "bad impression"), but who can blame them? Cash is heavy—$1 million in twenty-dollar bills weighs more than you can lift, and drug dealers have been disconcerted to note that their powdered merchandise is handier for smuggling than the equivalent money. Cash is inequitable—if you are one of the 50 million Americans poor enough to be "unbanked," you pay extortionate fees to seedy, bulletproofed check-cashing operations (even more extortionate than the fees charged for automatic teller machines, often up to 1 or 2 percent and rising). Cash is quaint, technologically speaking—unless you're impressed by intaglio-steel-plate-printed paper with embedded polyester strips (meant to inconvenience counterfeiters). Cash is expensive— tens of billions of dollars drain from the economy each year merely to pay for the printing, trucking, safekeeping, vending, collecting, counting, armored-guarding, and general care and feeding of our currency.

Cash is obsolete.

So here come Bitbux, E-Cash, Netchex, CyberCash, Netbills, and DigiCash—through the Patent and Trademark Office and into the marketplace. A frothing mix of public, private, semipublic, bank and nonbank institutions are rushing in with new forms of money. As the Internet booms, those experimenting with commerce at an electronic distance are struggling to perfect the sending of cash over wires.

The credit card companies have realized that their products are no longer about credit but rather about convenient payment for goods and services; they are entering the cash game with "smart cards" making heavily marketed debuts in Atlanta this month and New York at year's end. A British-based project called Mondex is promoting a global standard for digital cash at sharp odds with the Visa and MasterCard approach. Like their competitors, the Mondex people have noticed that cash changes hands 300 billion times a year in the United States alone, and they want their cut. They believe that electronic money has reached the stage of a classic emerging market and that the test for every participant will be to survive the next two years.

The big players are not alone. Internet start-up ventures, overseas telephone companies, universities, and city transit systems are all experimenting with digital payment schemes with extraterritorial ambitions. A battle has begun for market share—and also for a quintessential modern commodity, sometimes overlooked but always coveted: float. Float is wealth in transit—money that has been parked temporarily in a place where someone, probably not you, can earn interest on it. If the issuer of a traveler's check or subway token or smart card can grab a piece of your money and collect interest for days or even hours, it gains an edge. No wonder everyone seems to want to mint money— except the Mint, which is carefully standing aside, for now.

But are these new creations really money? When money cannot be touched, when it turns to electrons, when it dematerializes, some people start to worry about what they really have. Will it be enough for a bank, or a credit card company, or even the government to validate some chip as "money"—can it ever be as real as a dollar bill? The nightmare parable of digital cash goes this way:

You check your favorite leather jacket in a restaurant and get a receipt. On the way out, you present the receipt to the attendant. He sniffs it, rubs it, holds it to the light, cryptanalyzes it and—relief!—confirms its authenticity. He hands the receipt back and assures you: "That is your jacket."

Digital money is perfect money, flawless money, intangible money. It is money that has been robbed of its substance—the opportunity to get scuffed, worn, dirty, and perhaps lost. It is networked money, and point-of-sale money, and money on a card, and money on a computer. It is money that weighs nothing and moves at the speed of light. It is money incarnated, finally, as pure information.

That comes as a shock; yet information is what money has always been. It is information about value and wealth. Let's say you go online and buy a bottle of Martelli Vineyard Puncheon Select Gewürztraminer from Virtual Vineyards, an Internet site. The wine is real—the parcel service burns jet fuel to get it to your door. You are obliged to become $14 poorer and Virtual Vineyards is entitled to become $14 richer. You could send dollar bills by mail—those little green symbols of wealth acquired in past transactions. Once perhaps you could have sent a sliver of gold bullion—a symbol in its own way, now stripped of its special legal status. Only the information need change hands, so now you can send an experimental form of electronic payment called Cyber-Cash. Does it matter? This long-distance commerce becomes the ultimate extension of a process that began a century ago, when the Western Union Telegraph Company—a communications company, not a bank—jury-rigged a way to turn cash into bits flowing across wires. The wiring of money means something different now, when the transport medium, the Internet, is decentralized, international, and uncontrolled.

In a real sense, money has already gone digital—or virtual, or notational. The creased bills in your pocket are a language as outmoded as Morse code. The money supply of the United States amounts to more than $4 trillion. Every business day more than half that amount sloshes about among banks and other institutions in purely electronic form: signals flowing over wires. These accounts are reconciled by transfers not of actual dollar bills but of mere bits—the information is the be-all and end-all. "People today do not put $5 billion in a truck and drive it from one bank to another—that's just irrational," says Kawika Daguio, a specialist in payments technology for the American Bankers Association.

In fact, of the broad American money supply, only a small fraction—less than one-tenth—exists in the form of currency. All the bills and coins in consumer pockets, bank vaults, and elsewhere amount to about $400 billion. And most of that currency, as much as two-thirds, has long since departed the country, probably forever, mainly in the form of hundred-dollar bills. They belong to overseas money launderers and other enterprises that for one reason or another prefer not to keep their wealth in local denominations and local banks.

The vast daily traffic in money across the interbank network is not backed by dollar bills. Nor is it backed by the stores of gold at Fort Knox and elsewhere, the gold standard having long since gone the way of ducats and pieces of eight. "Money is the current liability of a bank," asserts Sholom Rosen, Citibank's electronic-cash guru. "It's as simple as that: it's not gold, it's not silver, it's the current liability of a bank."

You believe in banks, don't you? That's good, because ultimately money is backed by nothing but your own confidence, habit, and faith—a form of faith as powerful and essential to modern life as any religious belief. The coming digital era will make this plain to everyone, as never before. Still, the stock of old-fashioned cash out there is growing, not shrinking. About $1,400 in bills exists somewhere for every American. "That's a lot of paper money," says Lawrence Summers, the deputy Treasury secretary. "The question isn't why it's so small. The question is why it's so big."

Cash is growing, yet it is dying. You will carry some around for years to come, and perhaps barely notice when you stop using it in grocery stores, at gas stations, in vending machines. But the first forms of digital money to hit the market will not be the best forms; the rules that lawmakers have developed for managing paper money will not be the best rules.

For everyone who uses cash, everyone who stores it, and everyone who regulates it, a challenge is nearing. The challenge will be to make choices. Some kinds of electronic currency will protect privacy, and some will violate privacy. Some will make crime easier, and some will make it extraordinarily difficult. Some will tax commerce parasitically, and some will catalyze it. The new minters of money will have enormous power to choose—unless consumers, on the one hand, and government officials, on the other, decide to make their own choices.

"Digital *cash*—the stuff that circulates—isn't the only winner on the horizon, if it's a winner at all," says Daguio. Alternatives will be emerging from within the banking system and from the online world.

"If you didn't have to dig in your pocket for a coin and could drive by a tollgate at sixty-five miles an hour, everybody benefits from that," he says. "If you can send money to your children, or send money to someone you've never dealt with before, it opens up new opportunities and eliminates obstacles to electronic commerce. With credit cards there's always the risk that somebody isn't going to pay; with electronic cash that risk is done away with, and transaction costs can come down significantly.

"And people wouldn't have to pay those abominable fees to check-cashers. It's amazing what can happen, if the technologies are deployed correctly and the regulatory structure makes sense."

Vice presidents of Visa International, Visa U.S.A., and an assortment of associated banks have been thick on the ground in Atlanta over the past month, trying out the new cash card. It works in the Visa company cafeteria. MasterCard has already rolled out the equivalent in Australia and begun advertising it on American television. About 5,000 Atlanta merchants have agreed to install networked card-reading devices and accept the card in lieu of cash. Visa hopes to have "several million" cards in the marketplace before the Olympic Games end this summer—an ambitious number, considering that the Olympics expect to draw barely 2 million visitors.

The card is a chip embedded in plastic: a wafer-thin computer with, in this year's version, 2 kilobytes or 4 kilobytes of memory. The memory lets it store about eighty times as much information as the typical magnetic stripe on a credit card or fare card, and the processor makes possible the use of cryptographic methods to secure the data.

Using the card is supposed to be fast and easy. Unlike the cards that persuade automated teller machines to spit out cash, these smart cards require no PIN or password. Cardholders do not sign the cards; nor do they show any identification to merchants. They just insert the card into a small terminal. The card and terminal engage in a quick electronic conversation, validating each other's tiny identities, and if all goes well, a carefully recorded transaction takes place. The tally of cash on the card goes down, and the tally on the terminal goes up. If you lose the card, you have lost the money (don't come crawling to Visa—this is cash, or so the theory goes).

Visa and other credit card companies have adopted a technology that requires physical contact between card and card reader, but contactless technology is also available. You can wave your card in the vicinity of a turnstile or speed through a highway tollbooth, and a transaction can take place wirelessly. A card reader debits your card from a distance of a few inches or a few feet. (Convenient—then again, the next generation of pickpockets may need to do no more than brush on by you wearing the right card reader inside their raincoats.)

Considerable thought is being given to the question of where your cash goes—something like $2 trillion a year changes hands in amounts of ten dollars or less—and how to take slices of ever-smaller transactions. A Visa promotional video shows a motorist, having been caught speeding, cheerfully handing over his smart card to the trooper for instant justice: "Good afternoon, sir," the trooper says. "You have the option to take care of that right here on the spot." ("Is that his personal card reader or the county's?" Elliot Schwartz, analyzing digital cash for the Congressional Budget Office, says, laughing.) Children's allowances are a very real sliver of the money supply, and the Tooth Fairy is still believed to deal in cash. Mondex points out, euphemistically, that cash is "an important mechanism for spontaneous charitable donations"; it is certainly hard to imagine beggars trading cups for card readers.

At first the cards will be used until empty and thrown away; soon they will be reloadable. Next winter Visa and MasterCard plan a joint test in one of the world's most consumer-driven neighborhoods, the Upper West Side of Manhattan. The experiment is designed to make sure their competing cards will work in the same machines. And will New Yorkers use them? Smart card manufacturers know that many people dislike anything resembling computers. They know that cash has a magical aura for some—that in movies, for example, we love it (crassly) when bundles of cash glow from inside suitcases or piles of cash provide an aphrodisiac bedding material. Then again, in cooler moments we despise cash, too, so perhaps it is just as well that there is nothing romantic about a stored value card. Whatever the emotions at play, Visa cites studies suggesting that consumers tend to spend 5 or 25 or 40 percent more with a cash card than with cash—perhaps because they are lulled by the unreality of it all.

These spin-offs of credit cards will be the first big, mainstream digital cash, but this does not mean that they will succeed in dominating the market or—a separate issue—that they are ideal from the point of view of public policy. Still, the ultimate shape of electronic money will depend enormously on who wins the early market-share battles. Money, as a product, will offer a perfect example of the Law of Increasing Returns. The more people use any given type of money—the closer it comes to universal acceptance—the more useful and attractive it will become. Just like fax machines and Microsoft Windows, any particular form of electronic money will take off when, and only when, it achieves a certain level of penetration in the marketplace, a critical mass. By then it will have required a huge investment in the infrastructure of card readers and other associated technologies; that will raise the barrier for potential new competitors.

It seems natural to the credit card companies to divide the world between buyers and sellers. In their system, an extension of the credit card model, you are either a consumer or you are a merchant. Only merchants have the hardware and the authority to accept electronic money. Visa and MasterCard have made enormous investments in creating a payment-clearing infrastructure, all the millions of linked card-reading hardware around the world and the decades of consumer habits that make it all work, and their smart cards are meant to build on the power of that infrastructure.

The Mondex experiment is different. It imagines a world where everyone's telephone and everyone's computer can read money from smart cards and write money back. A trial has been running since last summer in the town of Swindon, England— Mondex cards loading and unloading in hundreds of stores and through street telephones with special screens. "I was able to give my six-year-old daughter a pound," says Tim Jones, Mondex's chief executive. "If we were both talking on Mondex phones, I could pop my card in, and you could pop your card in, and we could exchange funds. That for me is the core notion of a product that wants to call itself money."

Those hoping to ride the wave of commerce on the Internet agree that the distinction between merchant and customer is breaking down, along with the distinction between reader and publisher. Everyone on the World Wide Web suddenly seems to have information to sell, or advertising space. If only those Web-masters could conveniently collect a bit of your cash each time you drop by their sites! "It goes back to preindustrial society," Jones says. "Economies are brought alive by markets where everybody goes along as a producer of goods and everybody also goes along as a consumer. People can purchase information from their peers and sell information to their peers, just as if they were taking clothes or food to the agricultural market." With a difference: this marketplace is global.

These are expensive experiments—all those chips, all those terminals, all that marketing. The card-issuing companies presumably intend to make money, though they speak in slightly vague terms about just how. There are ways. They can take a slice of a few percentage points from every transaction, from the merchants' side. That way it is invisible to consumers; it is a kind of tax nonetheless. Issuers can sell advertising space on the card itself; residents of Singapore, for example, are already accustomed to using what look like miniature Calvin Klein billboards to pass through transit-system turnstiles. The credit card companies may be able to profit from information sifted from the vast mass of purchase records—information that could be of use in marketing. They may profit in a small way from one of the weird psychological side effects of anything to do with money: people like to collect it. Visa is already working with companies that specialize in marketing "commemorative" coinage, in hopes that customers will buy smart cards and set them aside, more or less forever, on the mantelpiece. And issuers of digital cash hope to profit generally from lost cards—telephone companies and transit systems already figure gains ranging from 1 percent to a phenomenal 10 percent—but there may be a surprise lurking in state escheatment laws, which require banks to turn over unclaimed accounts to the government after some period of time.

And then there is the little matter of float. "I've got a card for fifty dollars in my wallet, and it'll probably take me a month to use that up," says Doug King, a vice president of Wachovia Bank, one of the partners in the Visa project. "You multiply that out by millions of people, and there'll be some float there."

The ability of financial institutions to earn interest on your electronic money may not mean much now. When cash goes truly digital, it may mean everything. It is seldom recognized that the government benefits directly from the float on the cash in your pocket, the cash on your dresser, the cash waiting inside parking meters, the cash roaming around in armored trucks, the cash resting in the dachas of the Russian mafia, and all the rest of the $400 billion in cash outstanding. Holders of cash lend their wealth to the United States, interest free, just as holders of American Express traveler's checks lend their money to American Express. The Federal Reserve is required to buy and hold Treasury securities in an amount equal to that cash, and every year it turns over the interest it earns, currently about $20 billion. This income, known as seigniorage, represents revenue the government stands to lose as cash gives way to privately issued electronic currencies.

As, of course, it already is. At grocery stores and gas stations, ATM debit cards have an early head start; in Europe, gas stations now operate during some periods with a staff of zero. As cards wax, currency wanes: hundred-dollar bills are legal tender, in theory, but in real life many merchants will no longer take them—there is too much counterfeiting and too much plain uneasiness, as the Mint switches over to the odd-looking new-style bills. Certainly there is nothing to stop smart cards from replacing cash in stores, subway systems, and taxicabs, or at pay telephones and vending machines. Nothing, that is, but confusion, warring standards, business anarchy and, perhaps, a loss of faith.

Once upon a time, American commerce was bursting across a new frontier so explosively that the technology of cash could not keep up. Coins and bullion were too awkward to handle and too slow to move. In the early nineteenth century a multitude of banks, large and small, began issuing private notes instead: money made of paper. Immediately there were standards problems. Notes that were trusted in one state traded elsewhere at discounts that varied with distance, if the notes were accepted at all. By the outbreak of the Civil War, ten thousand brands of paper money were circulating, and as much of a third of it was phony. Only then did the federal government step in, creating a national paper currency and deliberately driving the competing forms of money out of existence through the imposition of a 10 percent tax.

This was controversial. The Supreme Court held that Congress has the power to restrain "the circulation as money of any notes not issued under its own authority." To a few officials, most notably the director of the Mint, Philip Diehl, there may be an interesting analogy here. "Coins are a declining second-wave technology of commerce," he has said. "What we are wrestling with here today are the implications of these emerging electronic third-wave substitutes for coinage." If smart cards are a new form of money, shouldn't they be issued by the one true minter of money, the authority with the power to cut through a war of confusing, conflicting standards—the government?

"Government-issued electronic currency would probably stem seigniorage losses and provide a riskless electronic payment product to consumers," Alan Blinder, then vice chairman of the Federal Reserve, told a recent congressional hearing. But he and most of the Federal Reserve and the Treasury have taken the view that direct government involvement in the roiling digital-cash business would be hazardous and stifling.

In the "current climate," as those in Washington tend to say, anything that smacks of an expanded role for the government is anathema. Policy makers at the Treasury are reluctant even to talk about electronic money on the record. "It's easy to go in and say, 'oh, we're going to regulate everything,' without knowing what everything is," says a senior Treasury official. "We want to know what everything is." He adds: "There are very serious policy issues—seigniorage, money laundering, financial stability issues, there are consumer issues that are genuinely important that we must address and look hard at. It may be sensible for the government to issue a card—that's conceivable—but what if you issue it and nobody uses it?"

A task force headed by the Controller of the Currency is considering these policy issues, at a deliberate crawl. The many private institutions getting into the money business agree, with all their hearts, that the government ought to just stay out of their way. They are not quite so worried that no one will use their products.

DigiCash, an Amsterdam-based company run by an American cryptography expert, David Chaum, is experimenting with money in varying degrees of reality. One version cannot be converted back to dollars or any other national currency, yet thousands of Internet users have begged to have some to spend in slightly whimsical Internet shops. The E-Cash Shop of Internet

Lining sells six online images of Japanese scenery, for one-and-a-half cyberbucks (c$1.5). For c$5, the American Book Center Grand Lottery Extravaganza will offer you the chance to win an actual, tangible, material object: a hardback copy of John Grisham's potboiler, *The Rainmaker*—"delivered to your home, for FREE!!" But real commerce is available, if not quite convenient, with another version of DigiCash's technology, issued through the Mark Twain Bank of St. Louis. This lets users store dollars in tightly encoded form on the hard disks of their computers. CyberCash, an altogether different operation based in California, issues digital "wallets"—items with no more or less tangible reality than the digital dollars they contain—for use in Internet commerce. At the moment, you have to open bank accounts with old-style cash, or at least turn over your credit card number, to get any of these digi-cyber-electro-dollars.

To bankers, this looks like anarchy. The one place they would like the government to take action is the place where nonbanks start to step on their toes: banks are subject to many regulations and safeguards that, so far, their less orthodox competitors remain gaily free from.

Over at Citicorp, the Emerging Technology group is creating (and fighting patent battles over) what the bankers hope will be the most securely based of all these systems. Their vision encompasses not just cash, but also the huge portion of the payments system that runs, almost as archaically, by check. Millions of people have begun paying bills electronically, tossing into the trash the little windowed envelopes that come each month. Nevertheless, nearly all company-to-company transactions today are by means of checks, expensive to process and highly vulnerable to counterfeiting. The average American signs 270 checks a year, according to Citibank, compared with ten for the average German. This is a burden of which Citibank would love to be relieved.

"We're going to have to go to this technology for reasons that have nothing to do with consumer convenience," says Colin Crook, senior technology officer of Citicorp. "This is profoundly important long term. It will change the entire infrastructure of banks. It takes $150 billion a year just to run the U.S. banking system. That's a crazy number at the end of the day."

Meanwhile, guises of money continue to multiply. The post office issues 200 million money orders a year. Traveler's checks, food stamps, even frequent-flyer miles are becoming tradeable and convertible to merchandise. Ersatz private monies have always existed—tokens, tickets, and chits of all kinds. But these are blending in the mainstream economy as never before, just as real cash comes to seem less and less distinctive. When so many objects can serve as money, government currency loses its special status.

Inflation and technology have conspired, anyway, to make American currency seem ill-configured as never before. The dollar bill of 1996 buys about what a dime did in 1941. The 1941 dime was more convenient. But the government has not seriously considered a dollar coin since the debacle of the poorly designed Susan B. Anthony dollar nearly two decades ago. Because we use paper money for amounts as small as the 1996 dollar, a whole vending-machine technology has sprung up to cope with dollar bills; still, your chances of straightening, uncrimping and stuffing any particular bill successfully into a machine are frustratingly small.

To return the currency to its 1940s condition, the smallest bill would have to be the ten; the smallest coin, the dime. At the bottom of the ladder, pennies are an expensive nuisance, blatantly disrespected. Take one, leave one. Billions of them simply vanish from the economy each year—another hidden cost of money. Oft-cited polls by Gallup and others that purport to show a continuing fondness for pennies—made mostly of zinc—are commissioned by, of course, the zinc industry. Many people do believe that eliminating pennies would lead to sneaky rounding-up price increases, but logic suggests this is not so: a two-dollar toy that now goes for $1.99 would likely drop to $1.90 rather than rise to $2.00. The Treasury believes officially that the currency is fine and popular as is. Americans are conservative about their dollars and cents. They may not want it to change, but it has changed and is changing: shrinking, fading, stepping back into a crowd.

Could a host of new monies undermine our collective confidence in Money—in the mass delusion that has made the United States dollar such a bedrock? Does the government have the responsibility, or even the standing, to take action? Federal officials are watching and waiting, hoping that dollars will always be dollars and trying to let many flowers bloom. "I think we should maintain an enormous presumption in favor of letting people participate in markets and compete and do all those things, while insisting on a whole set of regulatory safeguards that ensure that electronic money is not marketed by people who are then going to default," says Summers at the Treasury. Neither the government's traditional monopoly on the minting of money, nor the threat of lost revenue from seigniorage, persuade him that the government should act.

"I don't think that setting up the government Electronic Money Corporation is particularly attractive," he says. "That is the philosophy that brought you the world's state-run airlines, the world's state-run telephone companies, and the world's state-run electric companies, and by and large it hasn't been very successful."

True—but by standing aside, the government risks abdicating its responsibility for deep policy decisions. Consider, for example, the unanticipated rise of credit cards over the past two decades. Credit cards are no longer, for most of their users, a significant source of credit. They are simply payment devices— money at a distance. As a practical matter, it has become difficult to buy an airplane ticket or rent a car without a credit card. The vast bulk of mail-order and phone-order commerce depends on credit cards—or, more precisely, credit card numbers, for merchants no longer need see or touch the actual card. The credit card companies not only handle the payments conveniently; they have also come to serve, day in and day out, as a sort of shadow judicial system. You can trust a mail-order house with your credit card number as you would never trust it with cash, because you know that the credit card company will hear your complaints, examine parcel-service records, and back you if the merchandise fails to arrive. The days of C.O.D. are over. You pay for this system, of course, in the form of higher prices for everything sold by merchants who accept credit cards. In fact, because the credit card companies have mostly succeeded in forbidding merchants to offer discounts for cash purchases, you pay for this system even if you do not use credit cards—for example, if you are poor.

In effect, the economy has spawned an enormous privately managed payment system, financed by a hidden sales tax. A completely distinct, equally private system is the network of automated teller machines that has sprung up over the past two decades. These, too, carry high charges in percentage terms and are mostly unavailable to the poor. Electronic cash could evolve in the same way—public policy made without public debate.

In its quiet way, the government has contributed to the

decline of cash. Many people believe that, if they wanted to, they could get a thousand-dollar bill or even a ten-thousand-dollar bill. Not so—the United States has long since discontinued bills in denominations greater than a hundred dollars, even as it has added new laws making it harder and harder to make big payments in cash. It has imposed ever-tighter restrictions on your ability to drop off a secret suitcase stuffed with cash at your bank or at your lawyer's office. Cash, the government believes, has become largely a tool of criminals. This is true.

Whether or not you think you're ready for a smart card, surely you could be tempted by the not-so-smart card known in digital-money circles as the Evergreen Card. The Evergreen Card is digital cash with one flaw: the counter in its built-in chip gets stuck and fails to deduct the charge. In other words, it is the legendary magic purse, always ready with another coin.

Does it exist? Only in a cloud of speculation and myth, though a rumor swept a meeting of cryptography aficionados recently that Daguio, of the American Bankers Association, had actually seen one.

"Oh, God, I didn't say that, I really didn't," Daguio says. Not exactly, anyway. He merely pointed out that, no matter how good the design, technology always has implementation problems. Smart cards rely on software, and software always has bugs. "These things are so, so complex that just a little defect in a card in the wrong place on a random basis in a large batch could produce really interesting stuff," he says. "You know how hard it is to program software. Some of these bugs might not show up until you debit 60 cents, credit 40 cents, and then debit $1.50. If it becomes widely known that that's what you have to do . . ."

Visa officials are fairly sure that none of their Atlanta cards will be evergreen. "We'd never say never, but we haven't experienced that kind of a problem," says Gordon Howe, senior vice president. They also know that some people will not wait for the flaws but will try to create their own. "The chip is a physical thing; anything physical, the security people will tell you, can be attacked," Howe says.

As money enters a new age, so does counterfeiting. The ultimate threat is the perfect copy—the virtual coin that proves mathematically identical to the real thing. If money is a string of bits, then someone, somewhere, can make a perfect copy . . . and another . . . and another . . . An arms race is already raging between those working to armor-plate digital cash with doubly and triply secure cryptography and those working to pierce the armor. Security experts assume that nefarious characters, in search of an unending stream of money, are already investing millions in the next stages of research and development.

At first, issuers of smart cards with chips will be relying in part on how much easier it is to counterfeit smart cards with magnetic stripes. No sooner had New York's subway system virtualized its fare tokens in the form of magnetic-stripe cards than a few ingenious citizens discovered that they could throw together some cheap circuitry and heads from an old tape recorder and produce their own Metrocards. "You can reproduce cards ad nauseam—it's magnetic data," says Jerome Page, general counsel and vice president for business development at the Metropolitan Transportation Authority. At least the payoff is just a ride on the subway, for now, though the MTA, like so many other semipublic agencies, has wider electronic-money ambitions for its product. When counterfeits are detected, turnstiles can be reprogrammed by the central computers in a matter of hours, blocking the bad cards. Some users are shameless. "This is New York," Page says. "People come in to customer service and say, 'Joe sold me this $80 card for $30 and it doesn't work—I want my money back.'" The credit card companies, too, will be able to invalidate counterfeit or defective smart cards from a distance—if they detect them.

For every new idea in tamper-resistance, there is a new idea in tampering. Any chip can be cracked open and examined with an electron microscope—at least in theory. Manufacturers can put extra layers of oxide and metal over the silicon. Attackers need to etch these layers away carefully. Manufacturers can inject caustic agents that become active when exposed to oxygen, destroying the chip before it can be inspected. They can make some cells light-sensitive, programming them to erase critical data. Attackers can crack open cards in oxygen-free chambers or in the dark. Manufacturers can encase the chip in an epoxy containing diamond or carbide dust, to dull machine tools. "At least you can cause people to have to spend a lot of money," says Eric Hughes, a cryptography expert who is founder of Simple Access, an Internet services company. "But doing the second chip is far, far less money than the first. And if you could make a master chip that spoke the right protocol, you could make a little money mint for yourself."

He and others believe that the best strategy for a would-be latter-day counterfeiter is to work now, invest, and wait—resisting the temptation to attack the new systems in their embryonic stages. They also suspect that the biggest vulnerabilities will come from laziness and carelessness, not from inherent flaws in the technology. In the hot competition for early market share, companies may just cut corners.

"Information warfare is going to make people very worried downstream," says Crook at Citicorp. "We have an immense paranoia about how dangerous it's going to be. I think that the security requirements in our industry are going to be more severe than at the Department of Defense."

Follow the money, as Deep Throat said.

No one knows how much of the economy is still underground, in the sense of being untraced by banks, credit card companies, and Internal Revenue auditors. But if someone says, "You understand, I have a cash business," you do understand: a cash business today means an operation that evades taxes by concealment and deceit, keeping double sets of books and laundering money. If you learn that your city councilman has been putting cash into his safe deposit box, you are rightly hard-pressed to imagine a noncriminal explanation. If your painter demands payment in cash, you will not assume that he is paying his taxes in full. When small contractors routinely offer discounts for cash payment, it is not because cash is convenient, or because they enjoy the smell of old-fashioned greenbacks.

So a world where all money has gone digital could be a world where honest plumbers and restaurateurs, accounting for all their income and paying all their taxes, would not have to feel like schlemiels. Eliminating cash would free the world of the single biggest form of tax fraud. It could also wipe out a host of other crimes: bribery, kidnapping, extortion, and even robbery. All these depend on the existence of cash as an anonymous and untraceable means of payment.

Yet this could also be a world where vast computer databases keep track of every magazine you buy, every bus you ride, every hot dog you eat, every beer you drink, every video you rent, every sawbuck you borrow. If the network can follow the trail of all your spending, it can become more omniscient than a private detective who follows you around with a camera. In the money business, knowledge is power: your spending habits, your likes and dislikes, are valuable to marketers. That is not necessarily bad. Many smart card issuers have plans for including clever extraneous information on their chips: credit card companies imagine storing your vital health data, for example; transit systems imagine storing trip histories that would make possible frequent-rider discounts or special rates for bus-to-subway transfers. Still, the possibilities are chilling to anyone who cares about privacy. Separate information sources are becoming linked, letting your various watchers compare notes, and the detail in their dossiers is becoming finer and finer.

"The granularity of information that's revealed about payments is going to explode," says Chaum of DigiCash. In cryptographic circles, Chaum is the best-known advocate of a form of electronic money that could preserve anonymity, using advanced encoding techniques to create protective envelopes. Cryptography is as close as modern mathematics comes to magic. Banks could, for example, register every digital dollar they issue and verify the dollars when they return, while remaining unable to trace the spending trail in between. Chaum's scheme is not symmetrical: it preserves the anonymity of the buyer while recording the identity of the merchant. And by a clever mathematical device, if the buyer tries to spend the same digital dollar twice, his cover is blown.

He fervently believes that this is a flavor of money that consumers will want. "Privacy is inherent in the notion of a free market," he says. "If we don't get the national currencies in electronic form properly, then the market will route around them and make other currencies."

It's simply a design choice. Smart cards, or their online equivalents, could function as blindly as raw cash. They could be even less traceable than in Chaum's system. That is a frightening prospect to law-enforcement authorities. Having finally made life difficult for drug smugglers with heavy cash suitcases, they will not casually allow the manufacture of half-ounce chips that could make possible blind transfers of hundreds of millions of dollars: the money launderer's dream. Even if the government takes no other action in the electronic-money arena, it will surely move to extend its restrictions on cash to cover digital equivalents. And so far, the large institutions entering the electronic-money arena are leaning toward less anonymous, less private approaches than Chaum's—betting that most of us will be willing to sacrifice more pieces of privacy for, say, convenience. Chaum could prove right, but only if the marketplace is willing to cast its votes for privacy.

To a degree that is little appreciated, the government and financial institutions have already succeeded, mostly, in eliminating the anonymity of cash in bulk quantities. By making it difficult to move large sums of cash in secrecy, they have tightened a net. Most Americans will say in public-opinion surveys that they worry about their financial privacy, that they do not want to give anyone the ability to follow the cash trail they leave every day. Then, they go ahead and leave that trail without giving it a second thought.

Most of our economic life is already networked: every check, every credit card payment, every telephone call exists in a computer somewhere. Sure, you can sneak out anonymously and purchase a copy of *Penthouse,* but if you order a dirty pay-per-view movie from your cable company, you probably do not worry that the information finds its way to a database somewhere. Meanwhile, more than one adulterer has been revealed, more than one murderer hunted down, because they could not avoid leaving a trail of credit card receipts. If nothing else, the heightened fear of counterfeiting in the digital-cash world may drive banks, rightly or wrongly, to make sure that all money is networked money. "It's arguable," Daguio says dryly, "that banks have a right to determine whether somebody spent the same coin a thousand times." We have private money: cash. We have networked money: everything else. Digital money is inherently neither private nor networked. The technology can go either way. Ultimately someone will make a choice—the marketplace, or the government, or the credit card companies or the banks—and the technology will support it.

In 1996, virtually no one in or out of Washington thinks that the government should step in, take charge, and issue an electronic currency. There is a case to be made: That money is not just another product, best left to the vagaries of the market, but the irreplaceable underpinning of society. That confidence in money requires one currency, not a multitude. That only the government, after public debate among contending interests, can set standards equitably, rather than leaving the critical choices to, say, the credit card companies.

But the Treasury does not want the responsibility of guessing the future and maybe guessing wrong—creating an Edsel or Susan B. Anthony. It is just too soon. "There's this race," says Philip Webre, a principal analyst with the Congressional Budget Office. "We're at the gate. We don't even know how long the race is or how many horses are in the race." We do know that the computer industry does not want lawyers and congressmen imposing judgments about technologies of which they have proved famously ignorant. The old-line banking industry, and the nascent digital-cash industry, want the chance to sell their own products.

And, of course, they are all fighting for the float. Will you fight as hard for the right to earn interest on your cash? Will you decide that you want that last shred of privacy that comes with dollars that do not have your name and Social Security number built in? Or will someone decide for you? While you still can, why not reach into your pocket for a few last vestigial dollar bills, make sure you have exact change for the bus, and buy yourself a secret, non-networked hot dog.

OH-OH

June 1996

Prophets of doom come out of the woodwork at the end of the millennium, and this time most of the people forecasting a new reign of Satan, if not the apocalypse itself, seem to be managers of large computer systems.

They have noticed a little problem about dates. While most humans consider today to be Sunday, June 2, 1996, most computers save a few bytes by storing the date as **06-02-96**, or **960602**, or some equally pithy equivalent. Unfortunately, this means that after the celebrations end on **991231**, a new era will dawn with **000101**. To doomsayers, those zeroes look ominous.

Some are predicting that the opening seconds of the year 2000 will wreak a kind of havoc that would be exceeded only by the electromagnetic pulse of a nuclear attack: computers around the world crashing to a halt or, even worse, silently churning out miscalculated interest payments, lifetimes, annuities, and expiration dates. There are straight-faced forecasts of widespread business failures, a stock market crash, and a general depression with just one silver lining: skyrocketing salaries for computer programmers.

What is clear is that some computers will think it is 1900. If you have a typical PC, you could probably reset its clock right now to one minute before midnight, **12/31/99**, turn off the machine, turn it on again, and discover that you have gone backward in time to **1/4/80**, a strange starting date embedded in the original IBM personal computer. Programs running on most of the world's mainframe computers will be confused when they subtract one year from another—subtracting from zero may be perilous—in the process of calculating details like these:

- your age (18-year-olds may turn into minus-82-year-olds, or just plain 82-year-olds);
- the compounding debt on your latest credit card bill (99 years of 18 percent interest?);
- the shelf life of corned beef and prescription drugs;
- the locking status of bank vaults;
- the validity of your driver's license, health insurance, or latest paycheck.

A whole industry of Year 2000 specialists has burst into existence, including consulting firms that used to have more interesting, if less focused, missions. There are conferences and user groups. There is a newsletter, *Tick, Tick, Tick.* There is Wall Street fever for a few public companies with Year 2000 expertise. There is no shortage of work for the new millenarians, and some feel it is already too late.

"By the time you get into the '98–'99 time frame, you're better served to get into the funeral business," Kevin Schick, research director for the Gartner Group, says coolly. "There will be widespread panic by then."

This is a complex technology's way of sending us a delayed bill for the two bytes or two keystrokes saved every time, since the birth of computers, a database did not bother to store, or a person did not bother to type, that superfluous **19**. Some of the offending computer code has been running since the sixties, when few of the twenty-year-olds writing COBOL were in the habit of looking ahead a year, let alone a generation. In those days, when the primary input device was a punch card with eighty columns, every byte was expensive.

It's a quirk of numbers—an oversight or a necessity—and now it's time to pay up. The Gartner Group recently told Congress that the world will have to spend from $300 billion to $600 billion by 1999 just to get their dates straight—about as much as the United States will spend on gasoline in the same period. Federal agencies alone, they estimate, face a $30 billion cost—more than the total of the government's annual information-technology budgets.

Upon examination, these cost estimates are grotesquely crude. They are obtained by multiplying one huge uncertainty—the number of lines of computer code in existence—by another: the cost per line to fix bad code. The Gartner Group has computed the average yearly salary of a programmer, the time it takes to modify a line of code, the time it takes to test the modification, an "awareness cost," the cost of taking inventory of all that code, the cost of setting up work units, the cost of project management, and more. A lot of guesswork there—"and then I kind of added this fudge factor that says, over time you have the cost going up," says Schick.

So the numbers are soft and the rhetoric is Chicken Little-ish. Still, those who have actually plunged into old computer code and looked for quick remedies have been chastened. "The apocalyptic language that you hear?" says Capt. Don Brown, Year 2000 team leader for the air force. "Believe it." His group is just beginning, too late, he says, to inventory the "mission-critical" systems that will fail—"make sure our weapons are working and our planes will fly." That kind of thing.

The first government agency to begin grappling with the problem was the Social Security Administration, in 1989, when a program that helps recover overpayments tried to look a decade into the future and came to a dead stop. According to a joke that floated around in those days, every information-services manager in the world was planning a December 1999 retirement. Now the agency estimates that, with 30 million lines of code active at any given moment, it will spend 102 or 225 or 300 or 500 work years to forestall worse problems.

A federal task force is now seeking assessments from every agency and department. On the bright side, surely some of those billions of lines of ancient code running here and there will turn out not to do anything at all; they will be allowed to retire grace-fully. On the other hand, some software will turn out to be *very* expensive to fix, because the millennium problem is hard-wired in chips.

And in some odd way, it is also hard-wired in the culture of computing. You can't listen to Year 2000 specialists for long without realizing that part of the problem is fear: fear of looking stupid, and fear of telling the boss. The state of things as this millennium ends still includes a vast gulf in communications between the people who write code and the people whose companies live or die by that code. It's hard to explain the shortcuts, the compromises, the byte-by-byte trade-offs, the heads that bury themselves in the sand in the name of making a program run smoothly for a week or a decade.

And meanwhile, it's hard to explain that an issue so simple a ten-year-old could have foreseen it in the 1960s will now cost any medium-sized company tens of millions of dollars. Spent to gain what, exactly? IBM is circulating a 180-page white paper to its mainframe customers, advising, **"When you face your stockholders who will ask, 'Now that you spent XXX dollars on this project, what do we have now that we didn't have in 1995?' simply respond: 'WE STILL HAVE A VIABLE BUSINESS.'"**

BIG BROTHER IS US

September 1996

For much of the twentieth century, 1984 was a year that belonged to the future—a strange, gray future at that. Then it slid painlessly into the past, like any other year. Big Brother arrived and settled in, though not at all in the way George Orwell had imagined.

Underpinning Orwell's 1948 anti-utopia—with its corruption of language and history, its never-ending nuclear arms race and its totalitarianism of torture and brainwashing—was the utter annihilation of privacy. Its single technological innovation, in a world of pig iron and pneumatic tubes and broken elevators, was the telescreen, transmitting the intimate sights and sounds of every home to the Thought Police. BIG BROTHER IS WATCHING YOU. "You had to live—did live, from habit that became instinct—" Orwell wrote, "in the assumption that every sound you made was overheard, and, except in darkness, every movement scrutinized."

It has turned out differently. We have had to wait a bit longer for interactive appliances to arrive in our bedrooms. Our telescreens come with hundreds of channels but no hidden cameras. If you want a device with a microphone to record and transmit your voice, you are better off with a multimedia PC or, for that matter, a dedicated Internet connection: hook up your camera and turn on the switch that Winston Smith could never turn off. People in large numbers are doing just this: acting out their private lives before online cameras, accessible to the world. Grim though Orwell's vision was, it never encompassed the Dan-O-Cam (H. Dan Smith at work in his office in Fresno, California), the LivingRoomCam (watch children and pets at play: "personal publishing of personal spaces"), and scores of similar Internet "cams"—evidence that some citizens of the twenty-first century, anyway, will not be grieving over their loss of privacy.

And yet . . .

Information-gathering about individuals has reached an astounding level of completeness, if not actual malevolence. So has fear of information-gathering, if not among the broad public, at least among those who pay attention to privacy as an issue of law and technology. Hundreds of privacy organizations, newsletters, annual conferences, information clearinghouses, mailing lists, and Web sites have sprung into existence—a societal immune-system response.

The rapid rise of the Internet surpasses the grimmest forecasts of interconnectedness among all these computer dossiers. Yet it defies those forecasts as well. Strangely enough, the linking of computers has taken place democratically, even anarchically. Its rules and habits are emerging in the open light, rather than behind the closed doors of security agencies or corporate operations centers. It is clear that technology has the power not just to invade privacy but to protect it, through encryption, for example, which will be available to everyone, as soon as the government steps out of the way. The balance of power has already shifted from those who break codes—eavesdroppers and intelligence agencies—to those who wish to use them. In these closing years of the century, we are setting the laws and customs of a future built on networked communication, giant interlinked databases, electronic commerce, and digital cash. Historians will see our time as a time of transition. But transition to what?

"There's a very important and long-term debate taking place right now about technologies of privacy in the next century," says Marc Rotenberg, director of the Electronic Privacy Information Center in Washington. "Privacy will be to the information economy of the next century what consumer protection and environmental concerns have been to the industrial society of the twentieth century."

Privacy is a construct of our age. As a tradition in law, it is young. When Louis Brandeis issued his famous opinion in 1928 that privacy is "the right to be let alone—the most comprehensive of rights, and the right most valued by civilized men," he was looking to the future, because he was dissenting; the Supreme Court's majority was upholding the right of the police to tap telephone lines without warrants.

"In the beginning, there was no such thing as private life, no refuge from the public gaze and its ceaseless criticism," writes Theodore Zeldin, a social historian, in *An Intimate History of Humanity*. Then, he says, "the middle classes began cultivating secrets." In villages and small towns, the secret life was rare. The neighbors knew far more about one's intimacies, from breakfast habits to clandestine affairs, than in any city of the twentieth century. One's shield, if a shield was needed, was a formal civility: rules of discourse that discouraged questions about money or sex. The pathological case of the private person was the hermit—hermits, by and large, have disappeared. The word is quaint. In a crowd, we can all be hermits now.

"Privacy means seeing only people whom one chooses to see," adds Zeldin.

"The rest do not exist, except as ghosts or gods on television, the great protector of privacy."

In public opinion surveys, Americans always favor privacy. Then they turn around and sell it cheaply. Most vehemently oppose any suggestion of a national identification system yet volunteer their telephone numbers and mothers' maiden names and even—grudgingly or not—Social Security numbers to merchants bearing discounts or Web services offering "membership" privileges. For most, the abstract notion of privacy suggests a mystical, romantic, cowboy-era set of freedoms. Yet in the real world it boils down to matters of small convenience. Is privacy about government security agents decrypting your e-mail and then kicking down the front door with their jackboots? Or is it about telemarketers interrupting your supper with cold calls?

It depends. Mainly, of course, it depends on whether you live in a totalitarian or a free society. If the government is nefarious or unaccountable to individuals—or if you believe it is—the efficient ideal of easy-to-use, perfectly linked and comprehensive national databases must be frightening indeed. But if, deep down, you feel secure in your relations with the state, then perhaps you are willing to let your guard down: put off till tomorrow your acquisition of that encryption software, send your e-mail in the clear, perhaps even set up an Internet camera at the kitchen table or discuss your sexual history with Oprah.

Certainly where other people's privacy is concerned, we

seem willing to lower our standards. We have become a society with a cavernous appetite for news and gossip. Our era has replaced the tacit, eyes-averted civility of an earlier time with exhibitionism and prying. Even borderline public figures must get used to the nation's eyes in their bedrooms and pocketbooks. That's not Big Brother watching. It's us.

Like any gossip, we trade information to get information. Over in the advanced research laboratories of the consumer electronics companies, futurists are readying little boxes that they believe you would like to carry around—not just telephones but perfect two-way Internet-connected pocket pals. They could use Global Positioning System satellites so that you always know where you are. They could let the Network know too: then the Network could combine its knowledge of your block-by-block location and your customary 11 A.M. hankering for sushi to beam live restaurant guidance to your pocket pal. Surely you don't mind if the Network knows all this.

It knows much more, of course. Here is what exists about you in government and corporate computers, even if you are not a particularly active (or unlucky) participant in the wired and unwired economy:

- Your health history; your credit history; your marital history; your educational history; your employment history.
- The times and telephone numbers of every call you make and receive.
- The magazines you subscribe to and the books you borrow from the library.

- Your travel history: you can no longer travel by air without presenting photographic identification; in a world of electronic fare cards tracking frequent-traveler data, computers could list even your bus and subway rides.
- The trail of your cash withdrawals.
- All your purchases by credit card or check. In a not-so-distant future, when electronic cash becomes the rule, even the purchases you still make by bills and coins could be logged.
- What you eat. No sooner had supermarket scanners gone online—to speed checkout efficiency—than data began to be tracked for marketing purposes. Large chains now invite customers to link personal identifying information with the records of what they buy, in exchange for discount cards or other promotions.
- Your electronic mail and your telephone messages. If you use a computer at work, your employer has the legal right to look over your shoulder while you type. More and more companies are quietly spot-checking workers' e-mail and even voice mail. In theory—though rarely in practice—even an online service or private Internet service provider could monitor you. "Anyway," advises a Web site at, naturally, paranoia.com, "you should assume that everything you do online is monitored by your service provider."
- Where you go, what you see on the World Wide Web. Ordinarily Net exploring is an anonymous activity, but many information services ask users to identify themselves and even to provide telephone numbers and other personal information. Once a user does that, his or her

activity can be traced in surprising detail. Do you like country music? Were you thinking about taking a vacation in New Zealand? Were you perusing the erotic-books section of the online bookstore? Someone—some computer, anyway—probably already knows.

Many of these personal facts are innocuous in themselves. Some are essentially matters of public record. What matters is mere efficiency—linkage. Your birth certificate was never private; it was always available to someone willing to stand in line and pay a few dollars to a clerk at town hall. Computers and telephone lines make that a bit more convenient, that's all—but it turns out that proficiency in compilation, sorting, and distribution can give sinister overtones to even simple collections of names and addresses. A Los Angeles television reporter, to make a point, recently bought a list of 5,000 children, with ages, addresses and phone numbers, in the name of Richard Allen Davis, the convicted murderer of a twelve-year-old girl. The company that sold the list, Metromail, boasts of compiling consumer information on 90 percent of United States households.

To David Burnham, the former *New York Times* reporter who wrote the admonitory *Rise of the Computer State* more than a decade ago, this inexorably more detailed compiling of information about individuals amounted to one thing: surveillance. "The question looms before us," he wrote. "Can the United States continue to flourish and grow in an age when the physical movements, individual purchases, conversations and meetings of every citizen are constantly under surveillance by private companies and government agencies?" And he added, "Does not surveillance, even the innocent sort, gradually poison the soul of a nation?" Does it? If so, we're like sheep to the slaughter.

• • •

The right to be left alone—privacy on Brandeis's terms—is not exactly the same as the right to vanish, the right to act in society without leaving traces, and the right to assume a false identity. Most privacy experts who have studied the possible futures of electronic money favor versions that allow for the anonymity of cash rather than the traceability of checks and credit cards. That is appealing; we ought to be able to make a contribution to a dissident political organization without fear of exposure.

Still, the people with the greatest daily, practical need for untraceable cash are criminals: tax cheats, drug dealers, bribers, and extortionists.

Most drivers prove willing to use an electronic card to pass through tollbooths without worrying about whether a database is logging their movements. Yet if cards like these replaced cash altogether, the net around us would unquestionably be drawn a notch tighter—especially if we are lying to our employers or spouses about our whereabouts, or if we are simply planning to take it on the lam, Bonnie and Clyde–style.

In a past world of intimate small towns, people could disappear. The mere possibility was an essential aspect of privacy, in Rotenberg's view: "People left those small towns and re-emerged in other towns and created new identities." Could you disappear today: abandon all the computerized trappings of your identity, gather enough cash, vanish without leaving a trail and start life again? Probably not. Certainly, there have never been so many invisible chains to the life you now lead.

On the Internet, we are re-creating a small-town world, where people mingle and share news easily and informally. But this time it is just one town.

Some of its residents advocate rights not just to passive privacy, the right to be left alone, but to what might be called aggressive privacy: the right to retain anonymity even while acting with force and consequence on a broad public stage.

Passive privacy is the kind elegantly described by the Fourth Amendment: "the right of the people to be secure in their persons, houses, papers, and effects, against unreasonable searches and seizures." We do have a lot of papers and effects these days.

Aggressive privacy implies much more. Telephone regulatory commissions have listened to arguments that people have a right to remain anonymous, hiding their own numbers when placing telephone calls. On the Internet, surprising numbers of users insist on a right to hide behind false names while engaging in verbal harassment or slander.

The use of false online identities has emerged as a cultural phenomenon. Those who cannot reinvent a new self in real life can easily do so online. Sometimes they are experimenting with role playing. Most often, though, as a practical reality, the use of false identity on the Internet has an unsavory flavor: marketers sending junk mail from untraceable sources; speculators or corporate insiders trying to influence stock prices; people violating copyrights or engaging in character assassination.

Changing personae like clothing—is that what the demand for privacy will come to mean? It's a game for people who choose a form of existence impossible in the old world, maybe hermits at that, hiding in digitally equipped homes, visiting by telecam. Something has been lost after all, in the rush to modernity: the chance to mingle freely and thoughtlessly in our communities, exposing our faces and brushing hands with neighbors who know what we had for breakfast and will remember if we lie about it.

In compensation, our reach is thousands of times longer. We

meet people, form communities, make our voices heard with a freedom unimaginable to a small-towner of the last century. But we no longer board airplanes or enter schools and courthouses secure in our persons and effects; we submit, generally by choice, to the most intrusive of electronic searches. In banks, at toll-booths, in elevators, in doorways, alongside highways, near public telephones, we submit to what used to be called surveillance. In Orwell's country, thousands of closed-circuit cameras are trained on public streets—pan, zoom, and infrared. Every suit-case bomb in a public park brings more cameras and, perhaps, more digital hermits.

We turn those cameras on ourselves. Then we beg for more gossip. We invent diamond-hard technologies of encryption, but we rarely bother to use them.

If we want to live freely and privately in the interconnected world of the twenty-first century—and surely we do—perhaps above all we need a revival of the small-town civility of the nine-teenth century. Manners, not devices: sometimes it's just better not to ask, and better not to look.

FW: FWD>FWD- (FWD) A JOKE FOR Y

September 1996

FW: FWD>Fwd- (Fwd) a joke for y is the heading on this morning's e-mail. Something tells me I'm not the first person to read it.

This many-times-forwarded item is a case of electronic chain-letter humor. The current sample: a "Scientific Quiz to Determine Your Manliness Quotient"—a missive from Cynthia, who got it from Jane, who got it from Sunny, who got it who knows where and sent duplicates to who knows how many of her closest friends, each of whom in turn presumably sent it hurtling onward in bulk. ("**When is it okay to kiss another male? . . . When he is your brother and you are Al Pacino and this is the only really sportsmanlike way to let him know that, for business reasons, you have to have him killed.**")

Judging from the e-mail routing history and CC list—that stands for *carbon copy,* children—I am one of hundreds or thousands or tens of thousands of people reading the Manliness Quiz today. No problem. I fire back another item from my Inbox, **"Girlspeak-to-English Dictionary." ("She says: You have to learn to communicate. English: Just agree with me." "She says: It's your decision. English: The correct decision should be obvious by now."**)

Something is going on here that resembles the flow of water through rock. When rock contains relatively few microscopic cracks, water doesn't get very far. Add more cracks, though, and at a certain point—the point of percolation—they suddenly reach a level of interconnectedness that allows a plentiful free flow.

Your own little mailing lists of two or four or six sympathetic souls—hillary, al, dick, george . . . —form tiny enough pathways, but they start to interconnect. Where jokes are concerned, the online world is percolating.

Naturally we have jokes about sex, jokes about religion, jokes about lawyers, jokes about politics. Because of the still-skewed demographics of the Internet, we have rarefied and Byzantine jokes about computing and jokes about *Star Trek*. Much doggerel begins " 'Twas the Night Before . . ." or emulates Dr. Seuss. These categories are mixed and matched, as in, "If Dr. Seuss Wrote Star Trek: The Next Generation." Horribly morbid disaster jokes appear and spread (Chernobyl, Oklahoma City, and even Flight 800 humor) with the kind of astounding timeliness heretofore seen only in tightly knit joke-telling communities of cynical types with access to fast worldwide communication—namely, stockbrokers and journalists.

Joke-watchers may begin to sense humor surviving and evolving in Darwinian fashion. If recipients don't laugh, the jokes may not get passed on. Then again, it seems that nothing on the

Internet ever disappears, and one can always track down cultural favorites, archived somewhere. There are Letterman-style top-ten lists. In fact, there are Letterman top-ten lists, stolen and passed on, along with ever-popular Dave Barry columns. Originality is not a prerequisite. There are just plain lists of all kinds: Dan Quayle Quotations ("**Hawaii has always been a very pivotal role in the Pacific. It is IN the Pacific. It is a part of the United States that is an island that is right here**"); You Might Be a Physics Major If . . . ("**if you chuckle whenever anyone says centrifugal force**").

Jokes can slosh across cyberspace with a tidal force. One of last year's favorites began:

> **VATICAN CITY—In a joint press conference in St. Peter's Square this morning, Microsoft Corp. and the Vatican announced that the Redmond, Wash., software giant will acquire the Roman Catholic Church in exchange for an unspecified number of shares of Microsoft common stock. If the deal goes through, it will be the first time a computer software company has acquired a major world religion.**

Both Microsoft and the Associated Press felt compelled to issue press releases denying this.

Because e-mail is, sort of, written communication, long-winded text-oriented parodies seem overrepresented. But perhaps this new genre belongs to an oral tradition, too. "One way in which this stuff is oral, in a sense, is that it has the feel of folklore rather than professionally honed jokes," says David Feldman, a popular-culture expert and author. Once there was watercooler humor and barroom humor. Maybe e-mail chain humor is the ultimate extension—an electronic bonding experience. "Its charm is precisely that it is 'folk' humor rather than mass-mediated humor, akin to graffiti posted in an office that only coworkers would understand," Feldman says.

In older, slower, more fragmented times, there was traveling-salesman humor. "I suspect that the notion of traveling-salesman stories had as much to do with disseminating jokes from one isolated community to another as they did with the proverbial farmer's wife or daughter," says Kenneth Fisher, a New York City Council member and, evidently, humor theorist, who sends and receives more than his share. ("Believe me, I've only forwarded a small portion of the flow," he says. "One does not want to develop a reputation for telling bad jokes.")

Traveling salesmen are obsolete. These are fast and not-so-fragmented times. "I suspect that most of the circulators are also story-tellers face to face," says Fisher. "This just gives us more reach." Yes—much more. If the trend continues, it may soon be a matter of minutes from the time a joke is born to the time every human has received it.

HOLD THE SPAM
MAKE MONEY, NAKED BABES . . . ARGGH
December 1996

This just in from dreams@hotmail.com: "**You were referred to me as someone who may be interested in the following information.**" Oh, sure. I wasn't really "referred"; my e-mail address was harvested from Internet discussion groups or searches of Web sites or service-provider customer lists, and then sold and resold. A lot of my e-mail these days begins exactly the same way and continues: "**If you are not, please let us know, and we will promptly take your name off our mailing.**" This, too, is a lie.

Want to make money fast? Want to look at pictures of naked babes? Want FREE 1 yr. *USA Magazine* Subscriptions? A cure for heart disease? A $785,000 Dream Home Giveaway!!!? If so, then you are the person who is supposed to be getting all this mail. To judge strictly by my mailbox, the Internet, formerly thought to be a hothouse of intellectual and artistic creativity, has now mutated into a sales bazaar as scummy and senseless as any on the face of the planet.

The fellow at hotmail.com advised (with the odd grammar and capitalization that go with the genre) that "**If You Didn't Make \$5625.00 Last Week You Owe To Yourself And Your Family To Give our Program Serious Consideration!**" Just minutes before, a self-described M.D. with a return address of A-Winternational@msn.com asked me to imagine my daughter or son on the phone with a 911 dispatcher. "**Why? You are lying on the floor in the kitchen, clutching your chest, panic stricken. You are 45 years old. In addition to fear, lots of things are racing through your mind.**" You bet, doc.

One is that this is a hidden charge for all those free Internet sites that demand e-mail addresses and other personal information as the price of admission. Another is that humanity has never before encountered a form of advertising that costs its senders so little. Its targets, in fact, pay more, particularly if they belong to an online service that bills by the hour. Anyone with an Internet connection and a list of e-mail addresses can send millions of letters for, roughly, nothing. If you doubt that, just read my e-mail:

There's nothing that even comes close to this media of marketing. Everyday 10,080 new people log online that's 10,080 new prospects everyday! We'll put it to you like this lets say your selling a product for \$39.95 and you e-mail 1,000,000 people with your marketing letter and you only get a 1% response rate that's 10,000 ORDERS, you can do the math on this one!!

Are you wondering whether these marketers have any respect for privacy? They certainly do. Just not yours. "**There will be NO TRACE to your existing e-mail address**," they promise the would-be junk mailer. "**There will be NO WORRIES ANYMORE about sending out mass e-mail! With our service, you can mass e-mail till your heart's content!**"

Has a century of marketing science come to this? The promise, or threat, of the Internet is supposed to be that marketers will gather such detailed information about our personal habits that they will home in on us like laser-guided missiles. If they have something to sell to the fourteen people in the greater Kansas City area who like Thai food and drive a Chevrolet Suburban and listen to Crash Test Dummies, their databases will be ready. That is scary—but if they know so much about me, why do I keep getting mail from the online equivalent of dirty old men opening their raincoats: correspondents like jenny@babeview. com, whose epistolary method is to remark, "**WoW :{}**" and "**See ya**," by way of inviting me to look at fuzzy pictures of naked women?

Many Internet users object violently to mass mailings of this kind, calling them "spam" and pressuring service providers to cut off the offenders. America Online, whose mail programs deal with millions of such letters daily, has acted to block mail from a list of senders that it updates continually: from 1stamend.com and bulkemail.com to sweeties.com and youvegotmail.com, and including cyber-promo.com, cyberpromo.com, cyberpromotions.com, cyberpromo.com, cyberpromo.com, and cyberpromotions.com.

• • •

That last ever-mutating set belongs to Cyber Promotions Inc., a company that has been battling America Online and other services in court, so far unsuccessfully. "I can guarantee you that in a couple of years this will be a part of American life, just like every other kind of advertising," says its founder, Sanford Wallace. "I don't mind getting commercial e-mail at all, because I want to see what's going on out there. I also don't have a problem with watching frogs in the middle of the Super Bowl." He sees a bit of hypocrisy in America Online's response—back in the real world, America Online is famous for flooding nonvirtual mailboxes with promotional disks by the ton.

Mailings of paper and plastic, intrusive as they are, do cost money, so marketers have to exercise some judgment. The Internet makes it all too easy to fling random illiterate drivel across the planet, with fake return addresses. "I make a ton of money," says Frank de Roos 3d, whose company, de Roos Global Dynamics, is another big mass mailer (**"Business Opportunity. $10 000 Dollar Reward !!!"**; **"AT LAST!! Something That Will Positively Change Your Life Forever"**). He adds, "The largest demographic profession on the net is salespeople—you're talking about a bunch of salesmen selling a bunch of salesmen." Wallace's and de Roos's mailings typically come with boilerplate instructions for having your name removed from their lists—but carrying out such instructions is an exercise in futility, if my own experience is any guide.

Internet aficionados often note that the origin of the term

spam is obvious. Perhaps not quite. You need to recall the Monty Python skit about a couple sitting in a restaurant, trying to order some actual food, while a chorus of Vikings sings *"spam spam spam spam, lovely spam, wonderful spam, spam spam spam spam,"* louder and louder, until that is all anyone can hear.

A BUG BY ANY OTHER NAME

June 1997

"I don't know if I would use that word," a Microsoft support engineer said.

"What word," I replied innocently.

"You know—that three-letter word you just used."

Of course he wouldn't use it. He's under strict instructions never to say *bug* to a customer. In the official parlance of the world's most powerful software company, when a product is defective, one may speak delicately of an "issue." This could be a "known issue" or an "intermittent issue." Then again, it could be a "design side effect" or "undocumented behavior" or perhaps a "technical glitch." Excuse me while I go powder my nose.

Microsoft has developed a fascinating style of using language—Microspeak, let's call it—with a refinement, a subtlety, a fine polish all its own. To understand certain announcements and news releases issuing from Redmond, Washington, requires rhetorical analysis and possibly a glossary. The exercise can be worth the effort, though, because after all, Microsoft is Microsoft.

These days, if you don't savvy Microspeak, you're going to be left behind.

Consider a brief yet artful "Media Alert," released April 16 with the headline, "MSN Doubles E-Mail Capacity." (In slightly different form, the announcement appeared on Microsoft's Web site five days later.) It began, "Continuing to make the service more useful, MSN, The Microsoft Network, is doubling its number of e-mail servers."

Continuing to make the service . . . (A rule of Microspeak is, Never lead with bad news. The true reason for this media alert was that e-mail delivery to the Microsoft Network's 2.2 million customers had failed on a wide scale over the past weeks, reaching a point of crisis and forcing MSN to break its silence.) **. . . more useful . . .** (than what?) **. . . doubling its number of e-mail servers.** Of course, this is careful misdirection. The number of e-mail servers was not the real subject of the alert. The real subject made its first and last appearance in a subordinate clause in the following sentence: **Responding to a partial e-mail delay earlier this week . . .**

What is a partial e-mail delay? No further comment here, but the tail end of a second media alert three days later explained: **As MSN's membership grows and becomes more active, we process more and larger e-mail messages each day.** Translation: Our systems are overwhelmed by the volume of mail. **Earlier in the week we experienced some intermittent issues**—translation: breakdowns—**with our e-mail servers, resulting in delayed delivery of e-mail for some members . . .** delayed, if the truth be told, days and even weeks.

With further characteristic touches, the announcement also

noted: **No e-mail should have been lost** (delicious ambiguity there) **during this upgrade period** (don't forget, this is good news!). Emergency workers in the Mississippi flood plains, throwing sandbags onto the dikes, have encountered similar upgrade periods and intermittent issues.

In reality, according to users, considerable amounts of e-mail were lost and the problems have continued, more than a month after these two brief announcements—which are the only public statements Microsoft has made on the subject. Microspeak is language with a purpose, and it works, in a way. The press, both mainstream and technical, has just briefly noted MSN's e-mail troubles, in contrast to much more heavily publicized problems at America Online. And unlike America Online, MSN has not offered its users any refund for the lost service.

The odd thing is that people who work for Microsoft often have, as individuals, a candid, plain-speaking style. Most of them would be perfectly capable of saying: "We messed up. Here's what the problem is. We're sorry. We're going to try to fix it." Somehow the corporate culture has grown in a different direction. If Microsoft offers software labeled "preview," you may think you're getting a first look at a finished product. Actually, preview, in Microspeak, is what blunter software companies call "beta"— meaning incomplete, buggy, and unsupported. This spring, when the company rushed out a set of fixes for bugs in the mail software that shipped with Office 97, it dubbed these the "Internet Mail Enhancement Patch."

These are not lies, exactly, but they are a long way from truth. They are the form of debased language that George Orwell called "euphemism, question-begging and sheer cloudy vagueness."

"A mass of Latin words falls upon the facts like soft snow,

blurring the outlines and covering up all the details," he wrote. "The great enemy of clear language is insincerity." Lately we call this *spin*. Perhaps the Microsoft style seems all the more garish because this great institution operates not in the sphere of politics or war, where we've grown accustomed to pacification and collateral damage and plausible deniability, but in the technical realm, where words usually mean what they say. If what you want is a half round nose chisel or a dynamic link library entry point function, you had better call it by its right name.

Microsoft has brought spin "to a high art in the software industry," says Peter Deegan, editor of *Woody's Office Watch,* an online newsletter for Microsoft users. "The MSN e-mail debacle reminded me immediately of the story of how the old USSR is supposed to have announced the Chernobyl nuclear accident to the world media." Ah, Peter, if only. **Continuing to respond to users' desire for clean, inexpensive power, the Soviet Union has accelerated an upgrade of its historic Chernobyl plant . . .**

The company denies, by the way, that its technical-support people have formal instructions never to say "bug." However, the phrase "known issue" is preferred, a spokesman said, "due to the complex nature of the word 'bug.' "

This is what happens when you get too comfortable with Microspeak—you start to believe that *known issue* is simple while *bug* is complex. It is what Orwell saw as language of orthodoxy, of concealment, of the party line. He wrote, "A speaker who uses that kind of phraseology has gone some distance into turning himself into a machine."

THE DIGITAL ATTIC
ARE WE NOW AMNESIACS? OR PACKRATS?

April 1998

You probably haven't spent much time worrying about what will happen to your Web site when you're dead. That's all right. David Blatner is worrying for you. "I keep thinking," he says, "if my grandparents had built a Web site, wouldn't I want it archived and available on the Net in the years to come for their grandchildren?" So he is ready to help with his new Web-preservation organization in Seattle: Afterlife.

Meanwhile, in central Ohio, a site called Orphans of the Net is "rescuing" some Web pages that have been abandoned or shut down—generally shrines for minor celebrities. If you're looking for old publicity photos of Kimberly Williams or Renee Zellweger, rest assured that they have been preserved online.

These modest salvage jobs notwithstanding, many of the world's librarians, archivists, and Internet experts are warning that the record of our blooming digital culture is heading for oblivion, and fast. They note that we have already begun losing crucial scientific data and essential business records—stored on ancient punch cards or written in dead computer languages or encoded on decaying Univac Type II-A magnetic tape. (Just try to find a Univac tape reader when you need one.) In the electronic era, we are stockpiling our heritage on millions of floppy disks and hard drives and CD-ROMs. These flaky objects go obsolete dismayingly fast, with new technologies rolling in on product cycles as short as two to five years.

"There has never been a time of such drastic and irretrievable information loss," says Stewart Brand, creator of the *Whole Earth Catalog* a generation ago and an organizer of a sobering conference on Time and Bits. Our collective memory is already beginning to fade away, many of the participants believe. Future archeologists will find our pottery but not our e-mail. "We've turned into a total amnesiac," Brand says. "We do short-term memory, period."

The information-storage medium of the past couple of millennia—for words not writ in stone, anyway—has of course been paper. Paper does decay with time, and it is fragile. One big fire at the library at Alexandria in C.E. 391 destroyed a calamitous piece of the ancient world's heritage. But to some people, paper is beginning to look good.

"Paper at least degrades gracefully," says Brand nostalgically. "Digital files are utterly brittle; they're complexly immersed in a temporary collusion of a certain version of a certain application running on a certain version of a certain operating system in a certain generation of a certain box, and kept on a certain passing medium such as 5¼-inch floppy." If a company has digital business records a mere decade old, what are the chances that it has stored a vintage 1988 personal computer, DOS 2.1, and the correct version of Lotus 1-2-3?

Some companies have begun "refreshing" their aging records, by continually copying them onto new storage media using new software. Refreshing isn't easy, and most institutions have not yet realized that it may be necessary. Whatever media they used to save their digital information, they will not be able to read it without a machine—a finicky antique, most likely. With paper, all you need is your eyes.

Perhaps the speed and richness of the Internet have lulled us, letting children in Boise read Census data in Washington and oral history in Hiroshima. Words swim instantly across the network, not caring about the mileage, and we don't exactly feel information-deprived. We may be drowning, actually. But are we sacrificing longevity to gain glut?

"Back when information was hard to copy, people valued the copies and took care of them," says Danny Hillis, cofounder of Thinking Machines Corporation and now vice president of research at Disney. "Now copies are so common as to be considered worthless, and very little attention is given to preserving them over the long term."

It's scary. And yet . . .

Anyone wandering through the Internet might begin to feel that memory loss isn't the problem. Archivists are everywhere, in fact—official and self-made. On Sunday, July 3, 1994, I played a hand of bridge that would be best forgotten—but no, the leading online bridge service, OKBridge, has recorded every detail of the bidding and card play in each of the 7 million hands played since the beginning of that year. Likewise, any silly message that you broadcast to any Usenet newsgroup is now being stored, for eternity or some approximation thereof, by a variety of commercial services. No matter that you gave your last posting a mere five seconds' thought; you should be prepared to hear your biographer read it back to you in your dotage.

Most people, unfortunately, don't have posterity in mind when they fire off their little notes. Internet communication seems so spontaneous and personal. Will people really want future employers to dig up all the messages they've been posting to alt.dead.porn.stars and soc.support.depression.manic? Sometimes, as the years go by, privacy demands a gentle forgetfulness.

Many people sitting at company workstations toss off e-mail as casually as they speak—gossipy e-mail, secretive e-mail, snide e-mail, raunchy e-mail, e-mail meant to self-destruct after serving its instant purpose. But it lives on, as corporate lawyers and prosecutors have realized. Neither sender nor recipient can delete it reliably. To the lawyers' occasional horror—here comes the subpoena!—it lingers on disk drives and backup tapes like a late-night guest who has forgotten how to leave.

The biggest proprietor of archivable data is the federal government, struggling to preserve the records it generates daily on an uncountable scale. Literally uncountable: the last serious attempt was made early in the 1990s by the National Academy of Public Administration, which found—excluding the vast stockpiles of scientific data at the space and weather agencies, and excluding data on individual PCs—about 12,000 major databases; and the researchers also estimated that they had probably missed about the same number.

It is a matter of current litigation whether every piece of intra-agency e-mail must be preserved as a "federal record." Either way, the task is monumental. "What we're looking at is growth that there's no way we can deal with using any known technique or resources we can get," says Ken Thibodeau, director of the archive's Center for Electronic Records.

"Digital information technology is creating major and serious challenges for how we're going to preserve anything of our culture and our history," Thibodeau says. "It's also creating opportunities: we'll be able to preserve and use a lot more information than ever before." Pity the poor historian, though. The Clinton administration's e-mail for the Executive Office of the President alone figures to be 8 million files.

Meanwhile, in its unofficial way, the Internet is transforming the way information is stored. The traditional function of libraries, gathering books for permanent storage or one-at-a-time lending, has been thoroughly confused. Archiving of the online world is not centralized. The network distributes memory. There is a kind of self-replication at work, with data employing humans in the effort to spread and reproduce.

Web site by Web site, the data seem as frail as skywriting— smoke in the breeze. Brewster Kahle, inventor of some of the best Internet search systems, estimates the average lifetime of a Web page at seventy-five days. He has created an Internet Archive, though, to capture and store periodic snapshots of almost the entire World Wide Web. It saves pages that have been lost or shut down by their owners. It amounts to about 8 terabytes of data. ("Tera-" is 1,000,000,000,000. Get used to it. "Peta-" is coming.)

Brand and his fellow Cassandras have a point, and they are focusing attention on some new practical issues. Who, if anyone, will decide which parts of our culture are worth preserving for the hypothetical archeologists of the future? Can any identification scheme help readers distinguish true copies from false copies in the online world's hall of mirrors? What arrays of optical or magnetic disks might provide reliability and redundancy for more than a few years of storage? Still, hope comes from the simple truth that the essence of information does not lie in any technology, new or old. It's just bits, after all.

In the world before cyberspace, countless bridge hands were played and words spoken and the memory vanished like vapor into the air. Think of all that data, dissolving no sooner than it was formed. Once in a while people managed to snatch a bit back from the ether, with pen on paper or, later, audio- and videotape. They succeeded in saving for posterity a fair portion of what was worth saving: the speeches of Lincoln (the major ones), the poetry of Shakespeare (but not quite reliably), the plays of Sophocles (except the lost ones), and a few dozen terabytes more.

Everything is different now. The Internet turns a large fraction of humanity into a sort of giant organism—an intermittently connected information-gathering creature—and really, amnesia doesn't seem to be its fatal flaw. This new being just can't throw anything away. It is obsessive. It has forgotten that some baggage is better left behind. *Homo sapiens* has become a packrat.

Shed tears if you must for the backup tapes already demagnetized. You'll have many opportunities. Just last October, the Daioh Temple of Rinzai Zen Buddhism held a "Memorial Service for Lost Information," in Kyoto and online. Of course, the details are lovingly preserved, in English and Japanese, at its Web site.

CLICK OK TO AGREE

May 1998

I seem to remember that when I subscribed to, oh, the *New Yorker,* I sent in some money and eventually the magazine started arriving in my mailbox. Not so simple in the digital era. The other day I used my computer to subscribe to *Slate,* an online magazine owned by Microsoft, and after I gave up my name, e-mail address, postal address, credit card number, and choice of gift (I declined the free umbrella), the screen presented me with the first few lines of a 2,000-word contract. Below this was a button marked **I Agree**. There was also a button marked **Cancel**. I looked in vain for a button marked **Let's Negotiate—My Lawyer Will Be in Touch with Your Lawyer**.

I realize now that before you read any further we should agree on some ground rules.

First of all, by reading these words **you confirm your acceptance of, and agree to be bound by**, and promise never to call your lawyer to make light remarks about, **this Agreement**.

Furthermore, you're not buying a car or a toaster here. This department **makes no express or implied representations or warranties to you regarding the usability, condition, or operation thereof. We do not warrant that access or use will be uninterrupted or error-free or that we will meet any particular criteria of performance or quality.** No matter how bad the product is, it's your problem, not ours.

And after all this, if you think you've found a loophole and actually wish to sue, start by calling your travel agent, because **you consent to the exclusive jurisdiction and venue of courts in King County, Wash.** Oh, no, wait—that's Microsoft.

"Yes, it's absurd," says Michael Kinsley, *Slate*'s editor. But no more absurd, he adds, than agreements at other sites. Internet magazines are more complicated, interactive, and bug-prone than their print ancestors and thus require, in a litigious world, more complicated legal armor. You aren't really expected to understand it. "The entire software industry, for that matter, depends on its customers not really reading these things before clicking 'I accept,' " Kinsley says.

The software industry also relies on a clever legal twist: the notion that consumers are entering into ongoing licensing agreements with the manufacturers. You may think, as you walk out of a store, package under your arm, that you have bought that software. The industry claims that you have merely licensed certain limited rights to use it. It says so right there in the agreement you will find under the shrink-wrap and toss away unread.

As a licensee, you commit yourself to a set of continuing duties. In the case of *Slate*, for example, you agree to supervise any usage by minors and to notify Microsoft "promptly"—even though you've already paid—if you change your billing address, lose your credit card, or "become aware of a potential breach of security." Kinsley says he persuaded the lawyers to drop a clause that would have required all his readers to maintain their computer equipment in working order.

Are all these shrink-wrap and "click-wrap" agreements really enforceable? After all, customers have neither the time nor the expertise to read them, and often the agreements are hidden in boxes until well after the customers have paid up. No one knows for sure. In real life, manufacturers almost never try to enforce the sillier terms, and most of the damages people suffer from defective software tend to be in the nature of lost time—hours spent cursing the computer or waiting on hold for technical support—and it's hard to sue over that.

Steve Tapia, a Microsoft corporate attorney, says it just wouldn't be fair to hold software to the same standards as, say, a car. That's lucky for him, because carmakers have found it very expensive to sell cars with defects—especially defects they knew about. They can't just disclaim any obligation to guarantee their products. Software is different, Tapia says, "because personal computer software may be used for a myriad of different purposes on an infinite amount of hardware combinations."

In the early days of personal computers, users were mostly technical types willing to wrestle with flawed software. They forgave some of the bugs, in versions 1.0, anyway. Now that computers are a mass-market product, they reach more naïve customers who might actually expect their software to work. That must be why dozens of companies feel compelled to make users agree that they're on their own if they use the products in **hazardous environments requiring fail-safe performance, such as in the operation of nuclear facilities, aircraft navigation or aircraft communication systems, air traffic control, direct life support machines, or weapons systems, in which failure of the software could lead to death, personal injury, or severe physical or environmental damage.**

Some legal departments have been getting more creative lately. Customers who download Network Associates' antivirus software "agree"—click—to clauses designed to give the company control of press coverage: **The customer shall not disclose the results of any benchmark test to any third party without Network Associates' prior written approval;** and **the customer will not publish reviews of the product without prior consent**.

Meanwhile, the agreement that comes with Microsoft Agent, software that lets people create cute interactive animated figures, holds that you may not use the characters **to disparage Microsoft, its products or services**. Will the next version of Microsoft's operating system have a clause like that? I'll have to find a typewriter?

Perhaps some of these contract terms are striding defiantly past the limits of existing law—but the law is likely to change shortly, in all fifty states. A major revision is under way in the foundation of American commercial law, the Uniform Commercial Code. The drafters, a committee of lawyers established for the purpose, have created a new statute, Article 2B, specifically to cover software and other information products. To the horror of some consumer groups, the current draft ratifies the most aggressive provisions of today's software licenses.

It would set into law the idea that software customers aren't buying "goods" but merely licensing certain rights. It makes the licenses binding even when customers have not read them, when the customers casually clicked an online button, and when the customers could not have seen the agreements until after buying the products.

The draft legitimizes confidentiality and nondisclosure clauses like Network Associates', forbidding users to publish reviews of a product. And it would explicitly allow manufacturers to disclaim warranties; it even suggests language: **this [information] [computer program] is being provided with all faults, and the entire risk as to satisfactory quality, performance, accuracy, and effort is with the user.**

"It's the drafting committee's view that bugs are inevitable in software and that makes software different," says Cem Kaner, a lawyer and software consultant opposing these provisions. He argues that Article 2B in its present form will be a disaster not only for consumers but also for the more honorable software companies; it will reward companies that try to grab market share by rushing to market with buggy software.

"If there are no refund rights, no lawsuit rights, no legal disincentives, then companies that ship prematurely enjoy an unfair advantage," Kaner says. "In the process of protecting the worst companies from the consequences of their worst products, we pressure better companies to do a worse job."

CONTROL FREAKS

July 1998

A little window into Bill Gates's soul opened the day the Microsoft antitrust negotiations broke down for good. The light came from a plaintive comment by an anonymous Microsoft executive, who was trying to explain why it would just be so unfair if the government were to let computer makers substitute their own start-up screens for the Windows "desktop"—especially screens that gave users an alternative way to run their software.

"That would be tantamount to undermining our franchise," the executive told Steve Lohr of the *New York Times,* "by making Windows some bit of software plumbing that no one sees."

Once upon a time, of course, Microsoft was a little company whose only significant product was a bit of software plumbing that no one saw. At least, no one ever wanted to see it. This was the software called DOS, for Disk Operating System. An operating system was invisible plumbing by definition—the software that took the little signal coming from your keyboard, say, and sent it along as another little signal to display the letter A on your screen or to read program B from a floppy disk or to send document C to the printer. When you bought an IBM personal computer, you pretty much had to buy DOS with it.

Now it's fifteen years later and the operating system has gotten bigger and more complicated and more useful and more expensive. Microsoft's technical people think it's completely obvious that an operating system in 1998 should include Web-browsing services, because the need for these functions in our internetworked era is almost as clear as the need for disk-reading services. Those Justice Department lawyers far away in the benighted other Washington are so clueless, they think—especially Janet Reno, not even using a personal computer herself. Gates says that forcing Microsoft to remove these services from Windows would be like forcing Coke to remove ingredients from its supercool secret formula.

Don't we want Windows to be just as good as it can be?

Yes, we do! It really does make sense for the operating system to include all the core functionality of a Web browser: making connections across the Internet, interpreting the HTML code in which Web pages are written, displaying the results in a window. And all this should be built into Windows as plumbing—not as a single Microsoft-owned application, with every detail of the user interface chosen in Redmond, but as services that any software company can rely on.

Luckily for Microsoft, the appeals court ruling late last month on the first part of the government's antitrust case seems to give the company leeway to integrate just about anything it wants into its operating system—even, as the lone dissenting judge pointed out, a mouse. The two-judge majority looked through a microscope at the many possible shadings of the word *integrate,* but it averted its eyes from some side issues that will soon have glaring importance.

For example, if the point is to let Microsoft improve its operating system for the sake of the consumer . . . well, what is an operating system, anyway? It was supposed to be a foundation that supports all kinds of software, and that once meant that competing programmers should have equal access to this built-in Microsoft plumbing—at Netscape or anywhere else people have an idea for improving the Internet experience. You could imagine Microsoft's operating-system salespeople phoning Netscape to say, "Hey, heads up, we've got some great new Web-browsing functions coming and you'll want to use them in your next version." At least you could imagine that if Microsoft's real goal wasn't to suffocate Netscape.

Maybe these competitors would want to make a browser with a different kind of Search button. In older versions of Microsoft's Internet Explorer, users could customize the Search button so that it would go directly to Lycos or Yahoo or AltaVista or any other Web-searching site. Now, though, that Search button takes you to a Microsoft site, featuring a selection of Microsoft-approved search engines—search engines that have submitted to Microsoft contract terms for the privilege. The user cannot change it. Internet Explorer is most definitely not just plumbing. It's all too visible and in control.

Hundreds of thousands of exceedingly precise words remain to be written by lawyers pretending to fight and judges pretending to adjudicate the Browser Wars of the nineties, but those wars are over. It's impossible to imagine any judicial action that will remove Microsoft's browser from the desktops of virtually all the world's PC users. It's worth pondering a mystery that the court of appeals has not yet felt obliged to worry about. Why has

Microsoft spent so much money and muscle and corporate passion on a mission of giving away free software? You've got that Microsoft browser now. Why was it so important for you to have it?

The answer is that Microsoft is a forward-looking company with a sharp eye for the power points, the lever arms, the control valves in the emerging digital economy. In the software sector of the economy of the late eighties and early nineties, there was just one important power point: the operating system. Microsoft owned it and used it to gain dominance over that entire sector. In the helter-skelter Internet-driven world we're now entering, a variety of power points are taking form. The start-up screen, what you see when you turn on the computer, is a power point; Microsoft insists that manufacturers hand over all rights to that screen. *We're not just plumbing.*

Internet search sites are power points, because they can become habitual portals of entry for users seeking information or ways to spend their money. That's why it was important to make sure that your Search button would take you to a page at home.microsoft.com.

Before long, a power point will be the ownership of the standards for streaming audio and video data at high speeds across the Internet. A hodge-podge of companies have been working to create such standards—but Microsoft, proudly defending its right to innovate, has had its checkbook out, and it now owns all or part of each of these companies.

The Justice Department's lawyers understand this—the march of the Microsoft monopoly from one point to the next—but so far they have not found an antitrust framework that will let them craft a remedy. The most important power point of all is the very language of World Wide Web pages. This language has evolved rapidly, under the nominal control of an international standards-setting committee. Microsoft does not like international committees any more than it likes government lawyers, and when enough people are browsing the Internet with Microsoft software and nothing but Microsoft software, it will be time for that committee to pack its bags and go home.

WHAT THE BEEP

August 1998

The telephone rings. The microwave oven reports that food is ready. A clock announces the hour. Someone's pager has been paged. The fax machine has run out of paper. Or something—anything—has happened in a wristwatch, or computer, or alarm system, or dishwasher.

And you don't know which, or what, or whose, because all you've really heard is (pardon the transliteration): *beep*. Maybe it was *beep beep . . . beep beep*. If the television is on, you may not even be sure whether it was your phone or Jerry Seinfeld's. If you're at a business meeting, you may be one of a dozen mildly embarrassed people all reaching for your pockets at once.

Beep is just the beginning. We are surrounding ourselves with a cacophony of electronic sounds. Chirps, trills, hoots, shrieks—no, the old onomatopoeia can't capture these thin and perfect tones, and evolution doesn't seem to have prepared us for this particular assortment. They aren't the noises of the jungle. They're the noises of the human-factors laboratories.

At Timex, where they now sell "Beepwear" for the wrist, the engineers know they're limited by their tiny piezoelectric crystals, which vibrate in response to the tiny current from a very simple circuit. Wristwatch designers can only envy their counterparts at telephone companies, who can use little speakers and more complex circuits and make their pagers play "Hail to the Chief" when a message comes from the boss.

But musicianship isn't the goal anyway. "Everybody's beep sounds the same," says Phil Brzezinsky of Timex. "What we try to achieve in our products is, we want to make it as irritating as we possibly can, especially our wrist pager. We're always trying to crank up more sound."

A long, long time ago, telephones rang. They struck a pair of little metal gongs with a fast-moving hammer. This was the invention of Thomas J. Watson ("Mr. Watson, come here! I want you!"), improving on his and Alexander Graham Bell's previous noises, the "buzzer" and the "thumper." The distinct sound of the ringing telephone penetrated deeply into American consciousness for nearly a century, but in the spring of 1956, Bell Laboratories began field tests of a new sound, a "musical tone ringer," as it was optimistically called.

To smack those gongs with the little hammer took 85 volts; the transistorized tone ringer needed less than one. The three hundred subjects of the field test, in Crystal Lake, Illinois, mostly found the new sound "pleasant," after taking a week or so to get used to it. Strangely enough, an advantage was that they could easily distinguish their new telephones from ringing doorbells, alarm clocks, and fire alarms. In 1956, ovens, wristwatches, and the contents of people's pockets tended to remain silent.

●   ●   ●

Maybe now we are in the jungle, surrounded by all these new species and needful of more varied and sensible auditory cues. Even if we don't have perfect pitch, we're not completely tone deaf or rhythmically obtuse. If birders can learn to distinguish dozens of characteristic songs and telegraphers could handle Morse code, we should be able to cope with a few simple electronic warbles and trills.

"One of the advantages of auditory feedback is that it's 360 degrees," says Cynthia Sikora, a psychologist researching electronic sounds for Bell Labs, now the research arm of Lucent Technologies. "You don't have to be looking in a particular direction. It allows you to multitask and use your attention selectively. The problem with visual feedback is, you have to be attending."

On the other hand, our frequent confusion about electronic sounds suggests that they are somehow directionless. This turns out to be true, especially for the pure 2,000- or 4,000-hertz tones coming from a tiny vibrating crystal. A metal bell is a complex object, and it reverberates with a rich set of harmonics and overtones—and these, bouncing off walls, give a bit more information to a pair of human ears.

Several corporate research laboratories are devoting serious effort to creating different sounds for future telephones, computers, and other devices. Bell Labs, for example, is constantly bringing people in to test new sounds. Or is it the people being tested? Recently, members of groups were assigned different "personalized" rings and then asked to recognize them. "They were pretty good at doing some of them," says another researcher, Linda Roberts. "Other ones they couldn't tell." Even after going through the Familiarization Phase, the Learning Phase, and the Test Phase. Rhythms were the easiest.

Roberts and Sikora have been experimenting with what they

punningly call "earcons"—icons not for the eye. They have employed a professional sound designer and musician in New York to make the sounds brilliant and pleasant and identifiable. You might think in terms of (as they put it in a recent technical paper) "very short durations of program music that are created to steer the emotional reaction of the listener in support of the desired image." They have an almost infinite palette—but then again, sonic artists might feel constrained having to keep their creations shorter than two seconds, and often shorter than a half-second.

We navigate treacherous shoals here in the electronic world. At any moment you might be called upon to make an instantaneous decision: is it "join conference call" or "drop conference call"? Will you be able to identify the earcon ringing (and not exactly *ringing*) in your ears? Quick!

You can be sure that sometime soon, somewhere, an appliance will be speaking to you, and it will be saying, ". . . you will hear either a single beep, like this—*beep*—or two beeps, like this: *beep beep*." It will add, "After the beep, you may press *pound*."

ACCOUNTING FOR TASTE

October 1998

"**H**ello, James Gleick," said Amazon.com the other day (*click here* if you're someone else). "Take a peek at your brand-new music recommendations."

I peeked. Amazon's computers predicted that I would like the Beastie Boys, Adiemus, Frank Sinatra, Harvey Danger, and the Dave Matthews Band. What an impressive list! All right, I don't actually care for any of these, but still. It was quite a shot in the dark, considering I'd never been to Amazon's music department before. This is the way it's going on the Internet: if marketers want your money and your time and your "eyeballs," they feel they should figure out who you are and what you like.

Not only does their software try to calculate your taste in music by keeping track of the music you buy, it even tries to work out your taste in music from your reading habits. This could be a parlor game: If you like Vladimir Nabokov, maybe you'll also like Igor Stravinsky? If you like *War and Peace,* maybe you'll like the *1812 Overture?*

If you like E. L. Doctorow's *Ragtime,* maybe you'll like Scott Joplin's ragtime?

One Flew Over the Cuckoo's Nest and 10,000 Maniacs?

Consumer Reports and Crash Test Dummies?

Kafka's "Metamorphosis" and the Beatles?

We like to believe that our souls are our own and there's no accounting for taste. So it's disconcerting to find that, online, there's suddenly lots of accounting for taste. Amazon has its BookMatcher, the music store CDnow has its Album Advisor—sooner or later every merchant of just about everything will follow suit, analyzing your private likes and dislikes with "real-time recommendation engines" based on "collaborative filters" and fuzzy logic.

The basic idea is the same everywhere. Say you favor turtlenecks, convertibles, nautical history, bebop, and zinfandel. No doubt you are proud of your rare good taste, but it's a big world, and somewhere in a million-entry electronic-commerce database are a few other people with the same preferences.

If you knew that these people, your doppelgängers, were raving about the latest Mike Leigh movie, wouldn't you want to give it a look? In the jargon of the collaborative-filter game, these weird pals are your "community" and your "trusted associates." Their taste might be more in tune with yours than the few people you trust in your own small circle of friends.

The whole thing is just a mildly clever database look-up, but it works, at least for some people and some kinds of taste. It has no intelligence about the content of the merchandise. Mozart and Madonna might be flavors of ice cream, for all it knows. It merely has the beginnings of what could become a formidable electronic dossier: your purchasing history plus your volunteered comments about what you love and what you hate. At CDnow, for example, a customer can choose buttons for "I own this already" or "Not for me"; the computers, of course, watch and learn.

It's scary. "Is one's entire psyche's most secret landscape really a fairly public thing, given just a few pieces of information?" asks Douglas R. Hofstadter, the cognitive theorist and author of *Gödel, Escher, Bach.* "If you know that I love Chopin and Bach and am totally cool to Beethoven, can you predict that I love Cole Porter and Fats Waller but am indifferent to Oscar Peterson and Charlie Parker, and hate Elvis Presley?

"What is disturbing, to spell it out, is the idea that one's taste, which seems like such a personal thing, connected with and determined by one's inmost being, should have, in a way, a mechanically, nearly deterministically, knowable nature."

Maybe we're not quite knowable at that. And these are computers, so the mistakes they make can look very, very stupid. When they go off the rails in a sensitive area like taste, some people get angry. "The worst thing of all," says one irate customer, Russ Korins, "if you like Third Eye Blind—who sing 'Semi-Charmed Life,' that song that goes doo doo doooo, doo doo DOO do, and 'Graduate,' a mainstream version of Avenue A rebel rock, screaming, 'Can I graduate?!'—then what do they also recommend? *The Four Seasons* by A. Vivaldi."

At CDnow, they take this sort of thing in stride. "You can't argue with the customer—they know what they like," says Evan Schwartz, director of product management. "People love to click on 'Not for me,' 'Not for me.' "

When customers click, and especially when they buy, they add to the storehouse of information about that mercurial, irrational, chaotic thing we call taste. As knowledge builds up, maybe the computers will stop recommending Vivaldi to Third Eye Blind fans. Or maybe it will turn out that Vivaldi and Third Eye Blind have some kind of century-bridging affinity, even if no musicologist could say exactly what. Or maybe it's a moving target, and last year's Third Eye Blind fans have a different sensibility from this year's.

You might think of these growing databases as merchandising dossiers. Marketers are keeping a file on you, and if it's not as tangible or incriminating as your FBI file, it's too personal for comfort. "These exhaustive lists become much more than mere lists; they act as electronic psychoanalysts," writes David Shenk in his recent book, *Data Smog*.

Under pressure from privacy advocates, most companies in the collaborative-filtering business pledge not to share information without customers' consent. Even if the trail of your reading history leads your bookseller to conclude that you're on the verge of buying a new red Porsche Boxster or a blue Gap dress, Amazon promises not to tell the car companies or the special prosecutor.

Still, if we have learned anything, we know that information tends to get around. Do I really want the whole Web to know I'm a Beastie Boys, Frank Sinatra, Harvey Danger kind of guy? I didn't even know that myself.

PLEASE HOLD FOR MA BELL

November 1998

This being the nineties, I have a machine attached to my telephone line that answers the phone when I'm not there and records messages. (Let's call it an "answering machine.") So a woman identifying herself as a Bell Atlantic sales representative, "Lisa," is able to tell me that it is "imperative" I get in touch with her right away.

But how, Lisa? You work for that great black hole of modern communications, the telephone company.

She said she was in the Valhalla office, at 914-890-2550. I call. An automated voice-response system begins offering choices, all unappealing, none mentioning Lisa. Eventually I press 2 and "pound" and hear this familiar prose (surely it belongs in the next edition of Bartlett's):

"**All representatives are currently assisting other customers. Your call is important to us, so please hold for the**—" Then, silence. After a few minutes, I realize that I am engaged in the kind of psychology experiment where the longer you wait, the more you have invested in waiting, and so the longer you tend to wait.

With persistence and luck, I do eventually reach Lisa, and I ask her why, if she happens to know, the phone company's sales representatives are not equipped with two essentials of modern business life: their own telephone numbers, and voice mail.

That would be impossible, she says, because Bell Atlantic employs 9,000 people like her.

We live in an age when telephone technologies have transformed the relations between companies and their customers—and generally for the better. Why, then, do the telephone companies themselves seem to be the last holdouts? Why can't they use telephones and all the great new telephone accouterments like other businesses?

I respond to advertising and call 1-800-GET-ISDN, hoping, against my better judgment, to get a reasonably priced digital line for data at my home. For a second, it seems as though Bell Atlantic may have voice mail after all. "Your call has been transferred to an automated message system," a recorded voice announces. "The mailbox belonging to ISDN is full."

Sometimes a human being will give you a first name and sometimes a last name, but never both—contributing to a sense that the phone company is some kind of covert operation and you are the Enemy.

These are the time-honored habits of a monopolist. Despite the ubiquitous recorded voices saying "Thank you for calling Bell Atlantic," as though you had a choice, utility-style customer contact seems to be a cultural tradition dating back decades to before the antitrust breakup of the phone company. Where competition has arrived—in the long-distance business, for example, and the cellular-telephone business—human beings are a bit more like genuinely eager salespeople when they answer your calls for customer service. Old habits die hard, though.

At least the phone company now talks as though it faces the kinds of competitive pressures that nip at ordinary businesses. "We're facing such an explosion in applications for our services," says Bell Atlantic's CEO, Ivan Seidenberg. The company received 135 million calls last year, and he says it is highly focused on using the newest communications technologies to improve customer service—if you're tired of waiting on hold, maybe soon www.bellatlantic.com will solve your problem. "We don't think this journey is over with," he says.

For any company that receives millions of telephone calls, a key goal in this era of technological productivity is creating automated voice-response systems to handle as many calls as possible without any human intervention whatsoever. Designers hone and reorganize their menu structures, bringing in ordinary people to test them in the laboratory, watching to see if they will "bail out." Bailing out is giving up on the machinery—which you can do either by pressing 0 or just sitting on the line and pretending that you don't have Touch-Tone buttons at all.

Still, there is a gap between the theoretical picture in the command center and the reality at the front lines. In theory, the telephone company knows, or has the technology to know, the number you are calling from. It could have your file ready on a computer screen as soon as your call gets through to a person. This is called "screen pop." Bill Ball, executive director of systems and technology for consumer sales and service, says that Bell

Atlantic has already begun implementing screen pop and many other new technologies in the New York area. "We're in the throes of enhancing our system," he says.

But the reality is, for now, that you are likely to be asked to punch in your number by hand—and then, when you do reach a human, you are likely to be asked for your number yet again. After that, should you happen to be disconnected before your business is complete—and I am disconnected suspiciously often—you will have to start all over again and make your way back through the "vectoring system."

Why won't the person you were speaking with just pick up the phone and call you back? "We don't have time to make outgoing calls," a Miss Schultz tells me when I ask. "We're here for incoming calls." Ball, however, denies that customer-service people cannot call customers when they lose a connection. He asserts, "These people are very proactive relative to calling out." Oh, well. Anyway, Bell Atlantic is not alone. I get a strange bill from AT&T, demanding "final" payment on my "former" calling card account. This is all news to me (and wrong, as it later turns out). But when I call the designated number, a woman in the billing department tells me apologetically that the "computer" is "down," it has been down for at least a day, and can I please call back later in the week?

Well, okay, we won't even ask how the entire "computer" can be "down" at the once-proud American Telephone and Telegraph, but I'd rather she called me when it comes back up. No. Not possible. This is a telephone company on the planet Earth in 1998.

GREED DOT COM

December 1998

I have already mentioned a certain software company's using language in Orwellian fashion to obscure the truth rather than to reveal it, and I called this, with apologies to Orwell, "Micro-speak."

Soon afterward, I started getting e-mail from a Neal Sutz, of Tempe, Arizona, the new owner, as he explained, of the micro-speak.com Internet domain name. He wondered whether I might buy this hot property from him. The name, that is. I had better move fast, he implied, because he had faxed an offer to Bill Gates, too.

Well, golly, Neal. No!

But this just in from San Francisco: **Kevin Sinclair, owner of the domain name computer.com, is interested in selling the domain name. . . . Minimum cash bid is $500,000. If you are a start-up or an interesting public company, we are willing to listen to a proposal for stock.**

Computer.com and microspeak.com are two of the more than 3 million existing "dot com" domain names—bases for the e-mail and World Wide Web addresses that keep the Internet running. It's easy to grab a domain name. You would have to pay a mere $70 to Network Solutions Inc., the company that handles all this under a very sweet government contract, if you wanted to register, oh, say, gleick.com. (Yes, it's available—what slackers we are in my family!)

In 1994, when the Internet was young, it was even easier, and it was free. Practically the entire language was up for grabs, and Sinclair had the foresight or good fortune to sign up computer.com. But is he really entitled to $500,000 for something that costs approximately nothing? For a sort of high-tech vanity license plate?

"It's the free market," says Cheryl Regan, spokesman for Network Solutions. "It's a new world, you know, the Internet. There's a lot of rules that haven't been written yet."

It's not the free market at its prettiest. It's a few people trying to collect money not by actually contributing a product or service to the global economy—not by creating anything at all—but by snagging tiny bits of intellectual property that they hope someone else will want. Sometimes it works. Altavista Technologies Inc., which happened to have the altavista.com domain, sold it last August to Compaq for a reported $3.3 million—apparently the record. Compaq owns the popular Altavista search service—no connection to Altavista Technologies.

As the process has evolved, domain names that match well-established trademarks have generally been granted to the trademark holders. You know you can't go register coca-cola.com. But that leaves plenty of interesting, generic, unassigned words that have never before been thought of as anyone's private property.

Business.com went recently for $150,000. Cnet, an Internet company that already owned news.com, search.com, and shopper.com, offered a man named Mike O'Connor $50,000 for television.com. O'Connor, self-described "curmudgeon emeritus," turned the people at Cnet down. So they bought tv.com instead, for $15,000. They also happen to own computers.com (plural).

Then again, *own* may not be the right word. They have possession of these names, for now and for as long as millions of public and private computer owners, who collectively make up the Internet, continue to observe the convenient domain-name protocols that have worked for the last few years. Computers look up these names in tables, translate them into numbers, and use the numbers to route packets of data across the Internet. The system could change. It is certainly under stress—fast-growing and crowded. The federal government is determinedly getting out of the business of managing any part of the Internet. And the not-quite-official organizations taking over responsibility—most notably the new, nonprofit, international Internet Corporation for Assigned Names and Numbers—are struggling to find a way through the chaos.

Is this productive? Is it honorable? If you're not using a domain name yourself, what about just giving it back?

"Giving it back?" says Gary Kremen, the consultant who represents Sinclair in marketing computer.com. "It's property! That's like saying, I'm not going to use this piece of land, I'm going to give it back."

Sutz, with microspeak.com, feels the same way. He is a consultant in Tempe. He describes himself as a mere novice in the domain-name game. He spends a moderate amount of time trying to think of names that might have value and checking to see whether they are already registered, and he has signed up quite a few. He, too, compares domain names to real estate and says the "dot coms" are like Fifth Avenue addresses in New York. (Besides .com, there are so far .edu, .org, .net, .gov, .mil, and more than two hundred two-letter country codes.) Reserving these names is partly a matter of American patriotism, he says. Do we really want foreign businesses to have them?

So get 'em while they're hot.

Although I have definitely said no on microspeak.com, and Bill Gates seems to have other things on his mind, Sutz is not giving up. "I feel very, very strongly that it's only a matter of time," he says, "before somebody wants microspeak.com before their competitors get a hold of it."

MILLENNIUM MADNESS

January 1999

Historians of technology will look back at the Year 2000 computer problem—the Y2K "crisis," as newspapers now call it. They will draw from it a salutary lesson: We are a silly species, easily confused and given to sudden fits of hysteria.

At least we know how to make the most of a panic attack. Here's the latest issue of *Wallpaper,* a British high-style magazine, advising readers on how to "bunker down" for the coming apocalypse—with flashlights and gold bars and bottled water and, for self-defense, a stylish black crossbow. And why not? Supposedly careful authorities have raised the specter of power failures, plane crashes, food shortages, bank runs—the computerized wheels of commerce all grinding to a halt as the clock strikes twelve on December 31, 1999. In sober preparation for the Y2K crisis, the government of the United States has created more committees, action weeks, budget reports, presidential councils and orders, congressional acts, and, of course, Web pages than for AIDS and global warming combined—to name two genuine crises that actually do kill people and threaten the health of the planet.

As for us in the news media, suffice to say that we've gone mad—newsmagazine cover stories, special investigations, the works. We have been coconspirators of an army of people with a vested interested in seeing money spent on the problem: consultants (like the Gartner Group, quoted in almost every article), hardware manufacturers (none more alarmist than IBM), corporate and government information—technology managers, plaintiff's lawyers, and more. A confession: I committed one of the first *New York Times* accounts of the Year 2000 problem, back in June 1996. I'm sorry.

The amount of time that you, a dutiful citizen of the modern world, should spend worrying about Y2K is precisely zero. Do *not* call the new Federal Y2K Hotline. Do not call the manufacturer of your microwave oven. But if you care to devote three more minutes, you might consider the following:

Software has bugs. Already. Lots of bugs. My personal computer running Microsoft Windows crashes several times a week. Bugs have shut down telephone networks for hours, fouled up bills to millions of people, and caused at least one rocket to explode on takeoff. A mantra of customer service is, "Sorry, our computers are down." Somehow, dependent though we are on these finical machines, we muddle on.

Lots of Doomsdays. People focus on 1/1/2000. But computers have had to deal with the year 2000 for quite a while now. You probably already have a credit card with an expiration date of 00. Date arithmetic has always been arbitrary and flawed; all kinds of dates can cause trouble. Ready for 9/9/99? Doomsayers predicted trouble for 1/1/99, and technology reporters around the globe were on the lookout . . . and sure enough . . .

Redefining "news." January 1, 1999, produced some hot items that would not have been headline grabbers a year ago: Meters on 300 taxis in Singapore failed for hours beginning at noon; tragically, several riders were undercharged. In Sweden, drivers could not use credit cards at 600 gas stations because of a Y2K-like bug. And in hospitals, thousands of heart defibrillators made by Hewlett-Packard stopped displaying the correct date and time. They did continue to defibrillate, but doctors needing the time were forced to consult their watches.

The new McCarthyism. Virtually every computer-related company now shows off a "Year 2000 Compliance" statement, in accordance with the Year 2000 Readiness and Disclosure Act, a new law that, to oversimplify just slightly, requires everyone to swear that they take the issue very seriously. (Otherwise they might get sued someday.) In turn, companies require their suppliers and contractors to prove that they, too, are Y2K compliant. The Securities and Exchange Commission has mandated reporting and disclosure and testing and contingency planning by broker-dealers and investment advisers and public companies, and more. The thing is, it's hard to prove, but almost everyone and everything is Y2K compliant already. Sure, your car is riddled with computer chips, but they don't know or care what day it is. Personal computers really have no Year 2000 problems worth mentioning—and if you don't believe me, you can check right now by resetting your system date to 1/1/00. But no one can afford to look complacent. Even Microsoft has bent in the wind, feeling compelled to create the Microsoft Direct Access Year 2000 Readiness Disclosure and Resource Center—which notes shrewdly,

"Helping your customers solve their Year 2000 problems can mean revenue for you."

This prophecy is self-fulfilling. One alarming feature of the early Year 2000 warnings was that it would cost fantastic sums. This has already come true—never mind that much of the money has gone to the people sounding the alarm in the first place. That's not a disaster. Companies need to replace outdated computer software anyway, and the economy will somehow accommodate all the programmers who can now afford to fly first class. The one genuine risk confronting the world now is that people will take seriously the terrifying millennial forecasts and get even more panicky than they are: hoard cash and groceries, sell all their stocks, and arm themselves with crossbows or worse.

Look. We humans always fret about the end of the world when the calendar reaches a big round number. We're already anxious about computers, and rightly so: they're increasingly important, increasingly smart, increasingly interconnected, and almost as unreliable as ever. We live in a tightly knit world, where a failure in one place can have bad consequences far, far away. All this is true and, at the end of this year, some computers will have problems with their dates. But dawn will break on Saturday, January 1, and the western calendar will turn to the year 2000 C.E., and the sky will not fall.

FLOTSAM ON THE MOVE

May 1999

Stuff has been lying dormant in the world's attics and garages—stuff we might call "junk," like that early-American iron steam whistle, that 1938 Luzianne coffee tin (three inches tall and apparently used as a piggy bank), that 1962 Geiger counter with its original yellow carrying strap, and all the rest of the detritus of our productive little species. Suddenly this stuff is on the move. It's churning through the Internet's new auction bazaars like water through Hoover Dam, maybe just destined for future attics and garages, but providing considerable entertainment along the way. Economists have used the concept of velocity, V, to measure the turnover of the money supply. Now our flotsam and jetsam seem to be acquiring a velocity of their own.

Look at eBay, the biggest of the auction sites that have sprung up online. This place, if it is a place, is a billion-acre garage sale. One million, nine hundred ninety thousand, eight hundred sixty-six auctions are under way right now, in real time. More than seven hundred of them involve specifically Collectibles:

Advertising:Candy:M&M—boxes, cards, molds, planters, and a red diecast Reese's Pieces airplane bank ("only 5,000 were made," way back in 1992, the seller assures us). Pass them on. Then pass them on again.

It's not just old stuff, either. New, packaged, retail merchandise is flowing in along with the antiques and "collectibles," and it's all too easy to get hooked. I've never bought a laser pointer—I've never even *desired* a laser pointer—but it seems I can't resist. This is a "Featured Auction," happening before my eyes. The flashing blue icon says cool. The laser pointer is made of metal, not plastic, and the seller, one selwine@ix.netcom.com, stresses that it has been banned in many urban school systems. Also, it is vital in business presentations. Also, if you have pets, "they love to chase the little dot." But mainly I feel a hankering to own one of the century's essential pieces of technology, a powerful, pulsating energy machine that surely would have cost a lot more a few years ago. This is a Dutch auction, for merchandise in bulk as opposed to single items, and the ineffable Selwine calls for bids of at least $13.99. He promises that in six days, at precisely 8:19:17 A.M. PST, he will award a laser to each of the four hundred highest bidders. So, what to bid? I'm new to this auction thing, and I decide to type in a crafty "$14."

Then I wait. And browse a few of the other ongoing auctions, here in the land of exclamation points. The eBay Webmasters have organized and reorganized this cornucopia into "1,589 categories now!"—a giant directory, always in flux. Anyone on Earth, or at least anyone on the Internet, can sign up and list an item for sale, and then everyone else can start bidding. Clearly someone in Harrison, Maine, has run out of storage space for the May-June 1914 issue of *Bird-Lore* magazine, with articles on tropical bird voices and pictures of birdhouses made by young boys. No need to throw it away; have an auction instead. The process is meant to be easier than the average arcade game. You choose a suitably cute or mysterious user ID (**MUSCLES** or **BUYGONETRADING** or `lulu32`). You type in a few sentences of description (eBay encourages people to mind their spelling and capitalization) and, if you wish, put digitized photographs of the merchandise online. (This makes the erotica categories especially popular for browsing.) For a two-dollar fee you can add a bold-face title to your listing. For a dollar more you can have a **Great Gift** icon. For $99.95 you can buy a place in the rotating list of featured auctions. (You didn't think eBay was choosing these on the basis of merit, did you?) Sellers specify a minimum bid and a duration for the auction, and the eBay computers handle the rest. Thus, here is a **DROP DEAD**!!! one-of-a-kind Art Nouveau bronze chandelier with floral detail and mermaid cutouts, recently acquired from a Bel Air estate. I could bid on any one of the 4,645 Furbys for sale today. Or a car. Or a house. Or: **Soul of a 15-year-old girl, somewhat depleted and abused but still in good shape! Comes in an attractive cigarette box!**

The people at eBay aren't scrutinizing the offerings. They stay away from the actual transactions—shipping and handling, settling payment, that sort of thing. They're just lending us their virtual backyard for a vast, free-floating, never-ending tag sale. They estimate that more than a million different people connect to their site each day. More than $500 million worth of *stuff* got sold in the first quarter of this year, and the number has been almost doubling from quarter to quarter. Of course, they're taking a piece of the action: from $1\frac{1}{4}$ to 5 percent of every deal. Pretty much the entire human experience seems to be for sale, at eBay or some other Internet bazaar. Did you need "**Las Vegas Rms New Yrs Eve 1999 for 3 nights**"? How about a tee time for next Sunday morning—the 8:52 at the Heritage Palms Golf Club in Indio, California, is available—current bid, $91.

The new online retailers of books, recorded music, and computer equipment have invaded their respective business territories to much greater fanfare. And with reason—the flow of goods through the mercantile world has never before been rerouted so dramatically, so fast. The news that Amazon.com, formerly a bookstore, has bought chunks of Drugstore.com and Pets.com should make one wonder whether it was ever that much fun trudging to real-world shops to stand in line for vitamins, laxatives, condoms, or dog food. (And maybe the new online pharmacist will get prescriptions by e-mail, instead of spending his day listening to busy signals on doctors' phone lines.) But these retail enterprises, powerful though they will grow, are not the truest harbingers of the economy of the twenty-first century. That role falls to an assortment of odd new marketplaces—auction bazaars with names that drive copy editors mad: besides eBay, there are uBid, Onsale, AuctionBuy, AuctionInc., Auctionport, Auctionaddict.com, Bid4it, Up4Sale, BidAway, EZBid. Now Amazon, too, is starting an auction service and buying a company called LiveBid. Just as the World Wide Web seems to be turning every computer owner into a "content provider"—publisher, broadcaster, artist, commentator—the auction sites promise to make us all merchants. "We have to partner with thousands of sellers, even millions of sellers, over time," says Amazon's sportive founder and CEO, Jeff Bezos. "Our customers in a way are now able to compete with us."

Call these outfits junk mongers or facilitators of electronic commerce—either way, Wall Street is having trouble coming to grips with them. When eBay went public six months ago for $18 a share, *Forbes* magazine warned investors to watch out for the hype, the outrageous ratio of price to earnings, and the scent of "a bogus bill of goods." This was entirely reasonable. Three years earlier eBay had been nothing more than a personal World Wide Web page belonging to a bearded, bespectacled software engineer named Pierre Omidyar, who had also written a free program for playing chess online. Luckily for people who ignored *Forbes*, eBay's stock price has not only failed to collapse; it has climbed ten-fold. Now the financial markets assign to this little San Jose, California, company, with its hundred and sixty employees, a value greater than the combined worth of Kmart, United Air Lines, Apple Computer, and U.S. Steel. Unlike some Internet companies, eBay actually makes a profit. Anyway, the people driving the stock price aren't the pros; they're day traders, sitting at home computers, hopping in and out of the stock with an attention span measured in minutes. Their mental arithmetic goes, *Take all global commerce and multiply by 1¼ to 5 percent . . .* And when the market closes and they need to cool down, they point their computers to eBay and trade sports memorabilia (234,072 auctions running at this instant): trading cards, autographed balls, Olympic hand-me-downs. Maybe someone, somewhere, will actually bid $9.99 for this "**glass Snickers commemorative canister type candy jar**" from the 1992 games in Barcelona. It might have some kind of personal meaning.

The eBay people see their service as a counterforce to an overly commoditized world, with Gaps and Starbucks in every downtown. They talk about using technology in the cause of individualism—as though we were a race of people defined by our fetishes. Certainly the world has never seemed so full of tchotchkes and curios. Omidyar, who is thirty-one, started the company because his girlfriend collected Pez dispensers and needed an online forum for trading them. One eBay vice president collects Monopoly games and Depression glass; another collects hand-painted toy soldiers. "The ability to see that train set you got in 1956 that you're never going to see anywhere else, that's very powerful," says Steve Westly, the v.p. with toy soldiers. "People are at a certain fundamental level prewired to love collecting things. They want to return to their childhood or memories that have had a special meaning." Either that or we're big dumb squirrels who can't stop gathering nuts.

Executives at eBay and Amazon alike say that they are really building communities—many overlapping communities, composed of people who could never before have found so many secret fellow obsessives on the trail of Barbie shoes or Wonder Woman figurines. They have auction cheerleaders to egg us on. "**Express Yourself,**" urge the managers of eBay's jewelry neighborhood. "**If you could have any piece of jewelry that you wanted, money (and any other types of reality) being no object, what would it be and why? Click here to share your thoughts.**" Hundreds of people do click here:

Then let us dream. A diamond to rival the Culinan diamond set in the scepter of the Crown Jewels of England. A pink one, set in a brooch. Platinum setting. With another 10–15 carats of blue white diamonds (at least 1 carat each) for décor.

Sure, why not? Dear Santa . . . "**A single pink argyle diamond from Australia. Perfect.**" Or: "**Lots and** lots of Bakelite, I love it especially dots and gumdrop bracelets. But why stop there! I'd also love a deco platinum diamond, sapphire ring and a beautiful plique-à-jour (museum quality) Art Nouveau piece." And: "I think it would be great to have a piece of jewelry from everyone in the world so that you could spend every day looking at different jewelry. Now that isn't too much to ask, is it?"

No, it isn't too much to ask. It's an apt description of the world to which eBay and its many counterparts are pointing. People who care about nothing so much as jewelry can spend every day looking at different jewelry. They can buy, and sell, and buy, and sell, or at least bid on, or view pictures of, or dream of something not too far short of *jewelry from everyone in the world*. And likewise for people who prefer antique microscopes, Beanie babies, camera lenses, duck stamps, erotic photographs, or French brocade pillow covers. Or guns—eBay, reacting to criticism, has stopped providing a venue for the trading of firearms (including antique, collectible, and hunting weapons, as well as silencers, converters, high-capacity magazines, and ammunition with propellant), but scores of less fastidious sites have already leapt in to perform this service.

For humans as shoppers, barterers, traders, collectors, and hagglers, the online bazaars represent a rapid many-fold retreat of the horizon. In olden days someone in need of a teddy bear would explore, at most, as far as the handful of stores near home. The consumer's reach grew with mail-order catalogues and the

Yellow Pages (where fingers do the walking). We have telephone ordering and FedEx. Still, the options were limited. Suddenly the marketplace for teddy bears is truly global—not to mention the marketplace for Furby brand virtual pets. Not only buyers can reach farther; so can sellers. Even if you were the sort of person who liked holding garage sales, what were your chances of actually finding a buyer in Zanesville, Ohio, for those old enamel Bell Telephone cuff links? Now you just list them online with a "reserve" of $10 and see what happens.

Charles Chiarchiaro, a Maine antiques dealer who has been obsessed with old steam whistles since he was a boy of seven, now trades them on eBay, buying and selling simultaneously to a larger peer group than he ever imagined possible. Right now he's waiting to see if $1,100 will win him the variable-tone Crane Mockingbird fire whistle being auctioned by "Steamnut." He sees it as a "technological icon, a representation of technology and industrial art." Even when he travels to real flea markets, as he still does, he brings his PC along and monitors his eBay auctions through his car phone. I know how he feels. It turns out that I could have called Radio Shack or the Sharper Image and gotten a laser pointer for anywhere from $39 to $99, but that's a bit pricey. And besides, I hadn't realized I wanted one. Anyway, the Sharper Image has a new online auction site of its own. Meanwhile U.S. Pawn Inc., an operator of old-fashioned pawn stores, is rushing to open its own global online pawnshop cum auction service.

The auction process has its own entertainment value, in a loopy kind of way, even more hypnotic than the Home Shopping Network. It's a participant sport. Bonnie Burton, a compulsive shopper who has started an "eBay Weirdness Mailing List," says she is reliving her childhood in daily auctions of pop-culture toys—Little Miss No Name, Spirograph, Talking View-Master. She finds herself getting frenzied over a little waiflike plastic doll:

> I was willing to fight down online grand-
> mas for her. I started at twenty-five dol-
> lars, but ten minutes before the auction
> ended, I got outbid over and over again
> by some pesky old man collector. I wanted
> to e-mail him during the bidding to ask
> what the hell did an old man need a lit-
> tle dolly for? Was he a perv? Could I
> make him ashamed so he'd let me win? Yeah
> right. One thing you have to remember on
> eBay—no one has any shame, ever.

She hung in and won, for $60. In theory, bidders can avoid this kind of undignified last-minute craziness. You can tell eBay what you are "really" willing to pay, and the computers will increment your bid for you as necessary, keeping your bottom-line number secret from the other bidders. In reality, people seem to prefer making their heat-of-the-moment decisions, costly though that can be. "You start thinking of different ways to outsmart other bidders so you can get that mint-condition Super Spirograph," Burton says. "It's sick, and we're all happy to be addicts."

The average sale price for an item at eBay is a modest $46 or so. If I want to get more serious about this, maybe I should look at the online auctions under way over at Butterfield & Butterfield, the venerable San Francisco auction house, now trying, as they say, to "leverage our traditional model to the Internet world." Past meets present, and past keeps on meeting present. I could bid on first editions of Nabokov's *Lolita* and Madonna's *Sex* (the latter guaranteed to be unopened). Sotheby's is preparing to open a Sothebys.com auction business this summer. Christies.com hopes to start its Internet auctions in September, notwithstanding its scorn for certain unnamed "recent online auction initiatives." ("It is not surprising," declared Christie's CEO, Christopher Davidge, "that these initiatives have come under considerable criticism and raise concerns of potential fraud.")

All these houses stress their old-line virtues. They are the kind of auctioneers who appraise and authenticate and stand behind the merchandise that passes through their doors, even if the doors are virtual. Then again, Butterfield's just set a record, sure to be broken again soon, with an online bid of $138,000 for O. J. Simpson's Heisman Trophy in a real-time Internet auction. They are arranging modern-style strategic alliances, combining old players and new players with their assorted virtues: hence, auctions run in the name of Yahoo!-LiveBid-Butterfield, "Butterfield being the content partner, Yahoo! having all the eyeballs, and LiveBid having the technology," says George Noceti, Butterfield's senior vice president for Internet strategy. The people who really know auctions, he adds hopefully, are the old, old auction firms. Those mini-auctions over at eBay, with their low average tickets, just barely qualify as auctions. "They're glorified classified ads in most cases," Noceti says. "But the matching they do is the best I've ever see—people who want to buy and people who want to sell. That's one of the most compelling models on the Internet. Everybody can participate: you can put something on sale for a dollar; you can put something on sale for a hundred dollars." Or a thousand, or a hundred thousand. It hardly needs to be added that, after a hundred forty-three years in the business, Butterfield

& Butterfield filed for an IPO this year. And they have already canceled it. Instead, eBay is buying the whole company.

It's worth remembering that the auction and all its variants, including formal and informal haggling, predate fixed pricing in economic history. Academics have long argued about whether auctions—dynamic, raucous, equilibrium-seeking as they are—lead to fair prices. Why shouldn't they, as long as they are open and honest? Because we humans aren't robots. Richard H. Thaler, a University of Chicago economist, argues in his provocative book *The Winner's Curse* that people who win auctions almost automatically overpay. If you take a jar of coins to a bar, he says, and auction it off to the assembled patrons, the *average* bid will probably be less than the real value, but the *winning* bid will be greater—because everybody is guessing, and some people will guess too high.

You can make money this way. And if the online world is like a bar, it's very, very big and very, very crowded. Then again, not everything has a "true" value. It's easier to put a rational price on a jar of coins than on a vintage Little Miss No Name. "This has been debated by my peers," says Chiarchiaro, the steam whistle collector. "I have been told that prices are way inflated, hyped, and unrealistic. Wrong." Our obsessions and irrationalities aside, the global online auction does seem to represent a leap forward in price-setting efficiency, if only because of the sheer volume of information now sloshing around the world at high speed. It creates a vast, quick-changing library of price data. There are few secrets here. Thousands of laser pointers are up for sale on eBay at any given moment: *Liquidation! Blowout! Incredible closeout!* All these mercurial transactions establish a knowledge-driven market-place. In another eBay auction you can buy, for a dollar, a list of fifty wholesale laser-pointer distributors (delivery by e-mail).

So sellers like online auctions because they get a flood of new bidders and thus better prices for their merchandise. The bargain-crazed consumers who populate the auctions feel that they, too, are getting better prices. Can this be real? Kevin Kelly, author of *New Rules for the New Economy: Ten Radical Strategies for a Connected World,* sees a remarkable loaves-and-fishes thing happening here. "The miracle of both sellers and buyers doing better at once is really a statement about how clumsy and full of fat the usual way of buying something is," he says. "I don't think this paradoxical state where both buyer and seller gain together is sustainable in the long term. I think it is a sudden and temporary 'gift.' The efficiencies that the online auctions build on will quickly migrate to ordinary retail business, especially replenishables. The techniques used in auctions will benefit ordinary retail outlets; auction-ism itself will also migrate into retail, but in a not-in-your-face way. The bidding for things and the taking of bids will become so routine that they disappear into the background, done mostly by programs."

One tiny detail: payment is still not quite frictionless. At eBay, they just don't get involved, so, depending on the seller, people have to fuss with checks and money orders. Amazon, by contrast, is setting up a system using credit cards. At the less organized end of the spectrum, some people think that E-auctions will lead to E-barter. A person with, say, a surplus of Beanie babies and a need for outdoor furniture could find ways to trade and cut out the middleman. "Barter, sufficiently lubricated by electrons," says Kelly, "bypasses the inefficiencies in money itself." Maybe the whole mercantile economics of the past few millennia will turn out to have been a passing phase. People find themselves as addicted to selling junk as to buying it.

What's really flowing freely is, of course, information. We see now that high profits all along the distribution chain from producer to consumer have depended on pockets of darkness—places where information failed to penetrate. That's why local realtors and car dealers try to guard their price lists so jealously. It is why retailers have been slow to sell directly online, where they inevitably compete with their own distributors. It is why pure Information is a major eBay category: wholesale lists, how to make money on government CDs, Best Secret Info Sellers Don't Want U to See! ($9.95), and plenty of vaporous get-rich-quick schemes. In the auction bazaars of the Internet, there will always be someone with a lower price. "The world as we know it today is going to change," says George Noceti at Butterfield's. "I don't expect us to have new-car lots anymore. Everybody knows that you can get a better price online than at a dealer. I think what's going to happen is that the manufacturers will have sites where they demo cars and where they repair them, but when it actually comes to buying them—you don't care where you get it." The auction model has begun to transform the business-to-business supply chain as well. Industrial procurement of coal, machine parts, plastic moldings, cosmetics ingredients, and specialty steel has all turned to real-time, interactive, online bidding, saving money and cutting out middlemen.

Jeff Bezos, at Amazon, doesn't think electronic commerce means the end of stores, any more than television meant the end of cinema. He finds himself in the enviable position of being able to condescend to what we used to call *reality*. "The fact of the matter is, the physical world is the best medium ever," he says. "It's an amazing medium. You can do more in the physical world than you can do anywhere else. I love the physical world!" Hey, don't we all. But it's 8:19:17, and, lucky me, I have won a laser pointer. (In E-auction-world, we don't buy things; we win them.)

Selwine@ix.netcom.com steps out from behind the curtain and reveals himself as Spencer Zink, a sometime wine salesman from Torrance, California. Only thirty-six people bid on this lot of several hundred laser pointers, so we all win, and under the rules of the Dutch auction we all pay the same as the lowest winning bidder—in this case, the original $13.99. Plus shipping and handling, of course. We're inclined to trust Zink and cheerfully mail off our personal checks, because he rates well in eBay's feedback system, where we users can both praise and slander our fellows, for all to see. Talk is cheap, and there are plenty of shady buyers and sellers trying out various scams online. Consumer-fraud bureaus in many localities have been struggling to find a way of dealing with the eBay phenomenon, but it isn't easy, because after all, they are in *localities*. Anyway, almost a thousand of Zink's past buyers have posted comments like, "Great seller, excellent deal, super fast shipping—I'll bid again!" He started about a year ago with some wine he happened to have around. Then he looked around, "did some homework," and found cheap sources of merchandise like laser pointers, wholesale, from Taiwan. Sometimes he uses the more traditional auction style, starting off with a minimum bid as low as $1. He tries to double his money, while undercutting the retailers by 50 percent or more. He has sold about 4,000 items in this manner, and he is a piker.

So back to eBay's home page. A genuine 500-billion dinar note from the former Yugoslavia can be yours for $9.95. (Eleven zeroes!

Be a billionaire!) If you're an adult, or you like pretending to be an adult, you can bid on pornographic pictures from all eras, in all media, not to mention leather whips chains handcuffs vinyl, and more mysterious technologies. Several dozen Internet users in the United States and Canada are vying for a few cases of military-surplus meals—chili with beans, chunky beef stew, five-year-shelf life, "a must for those preparing for Y2K." And, lo and behold, Spencer Zink has a new auction under way. It seems he has some laser pointers to sell.

MULTIPLE PERSONALITY DISORDER

December 1999

Who can you trust online these days? Not me. I seem to have posted Nat King Cole's classic rendition of "Naughty Angeline" online for the whole world to download, free. It's right there in the alt.binaries.sounds.78rpm–era newsgroup, a forum filled not with messages but with mostly illicit copies of recorded music, digitized in the popular format known as mp3. People are e-mailing me to ask for more. "**Thanx for the great posting!**" says msmadge@home.com. "**Could you please repost this song as it was incomplete. Thanx again!**" You're welcome, Madge! I feel so generous. And yet so naughty, because after all, this music is protected by copyright, and Blue Note Records tries to sell it for money.

It turns out that the perpetrator of this particular crime was not really me. I'm just getting all his mail, because he made up a fake Internet identity—skink@around.com—that happens to point to me. There are words for this. "It appears you have an mp3 poster who has co-opted your domain," explains David Ritz, an expert volunteer monitoring abuse of this kind. Domain names are essential pieces of e-mail addresses, like town names in old-style addresses. They're the piece that comes after the @ sign. "Whitehouse.com" is a nice, recognizable domain name. (It is a pornography site, but that's another problem.)

When I signed up long ago for "around.com," I casually arranged for all around.com mail to come my way. All of it—from a@around.com to z@around.com. I failed to foresee that an easy word like "around" would be an accidental magnet for people making up false e-mail addresses. "Skink@around.com" evidently just popped into someone's head. Skink has plenty of company. I get mail meant for scores of creative souls who spontaneously chose to call themselves nowhere@around.com, look@around.com, kid@around.com, fooling@around.com, run@around.com, dick@around.com, clown@around.com, screw@around.com, slap@around.com, sleep@around.com, sniffin@around.com, horsing@around.com, going@around.com, andwesat@around.com, cu@around.com, idontleavethislying@around.com, and worse.

I feel I know them. Skink's musical taste runs to forties blues and modern hammered dulcimer, so we can call him eclectic. Lately he's been disseminating the works of T-Bone Walker and Jim Fyhrie. He also likes golf, if I can judge by his incoming junk mail. This junk is as perspicacious and misguided as only Internet marketing can be. "Forgive the intrution but I know you love music," chirps someone from British Underground Rock Bands in a typical missive. I can't stop myself from reading it. Skink is somehow my e-mail cousin.

My in box is getting a picture of how people behave when they pretend to be someone else, and it isn't pretty. At best, the online world looks short on faith. On bad days it feels like a grotesque masquerade party—all visitors in clownish disguise. Discussion groups feature angry, suspicious exchanges among people all employing pseudonyms. Here is Bulldog (sniffin@around.com) trying to shout, "**I don't trust you,**" at someone called Igetitnow@wow.com, participating in a lunatic-fringe bulletin board devoted to Year 2000 survivalism. "**Are you not the same 'Rooster Cogburn' that just a week ago 'got it'? I think you want people to send you money in the hopes that they can come and 'camp out' and share your supplies. I think you are a fake.**" He wants this invective to hit his target, but no, it just bounces from one false address to the next.

"**Sorry, I couldn't find any host named wow. com,**" writes a computer redirecting misaddressed mail at MIT. "**This is a permanent error; I've given up. Sorry it didn't work out.**" It's sad that this robot is the only polite and dependable speaker in the whole bunch, but it's no accident. There's a reason: the machine is the one identifying itself with an honest name and address.

Many of us, when we first get online, feel the thrill of being able to play the game without revealing our whole selves. "On the Internet, nobody knows you're a dog," said Peter Steiner's classic *New Yorker* cartoon. Our every instinct tells us to protect our privacy. If we don't think carefully, we're tempted to imagine that privacy and anonymity are the same. Our names have value, after all. Marketers covet them. They list us in their databases and sell our contact information and intimate shopping preferences to the highest bidder. So nonstop hit-and-run skirmishing has broken out between the forces of imperialism and rebellion. We natives try to raid their malls and information services and porn sites, grab what we can and slip away again under cover of darkness. How can they be sure we're who we say we are? Authentication is hard.

Computer scientists and cryptographers have invented schemes for proving identity with digital "signatures," clever code attached to e-mail messages, but hardly anyone uses these. The most common real-life technique for authenticating an online visitor requires just a few simple steps. A visitor needs a password to enter a site with Web browsing software. The site demands an e-mail address. The visitor hands it over. The site sends the password, not back to the browser, but by e-mail. If the e-mail address is genuine, the transaction succeeds. But if the e-mail address is someone else's, then a total stranger will get an unexpected password, with some helpful explanation like the following (from "CmdrTaco" of Slashdot.org):

> `If you didn't ask for this, don't get`
> `your panties all in a knot. You are see-`
> `ing this message, not "them." So if you`
> `can't be trusted with your own password,`
> `we might have an issue, otherwise, you`
> `can just disregard this message.`

You'd think Internet users would see how pointless it is to have their essential security information forwarded to fake addresses, where they themselves will never receive it. Apparently not. So their passwords come my way. I have a password for someone called "Bite Me Again," which enables me to download a demo of the Cinepak AVI Codec, as soon as I figure out what it is. I have a password that gets me past the bouncer at Kelly's Barroom, which seems to be a sad sort of chat site. I am identified as Chunky Monkey. "**I'm here, let's party**," I seem to have announced, willy-nilly. Rick, too, is saying, "**I'm here, let's party**." Sara C S retorts, "**Anybody online?**" and then, "**We gone, bye bye.**" The Algonquin Round Table it isn't.

Then I get someone's password for Live Fucking Uncensored and Explicit, "**the nets most shocking explicit site!**" I could watch their Live Teens and Live Sluts, their Shower Cams, College Girl Cams, Dressing Room Cams, Toilet Bowl Cams, Upskirt Cams, Gyno Cams, and cams that are even less respectful of the modern woman. As in so many areas of human endeavor, interactivity is now the watchword: "**You pick the TOY. You pick the GIRL. You TELL her where to stick it.**"

The pornographers want my credit card number, of course. So do legions of scam artists who are sending their hopeful missives to my unreal addresses. "**You posted me a note saying you wanted to make money using the Internet but you were not sure how,**" writes one, trying to communicate with a fake address by using a fake address of his own. "**All we do is help people find what they are looking for. We have helped a lot of people over the last few years. If you want the help email me your complete mailing address. Also, email your complete phone number. I want to talk to you. I nor any of my Consultants mess around. We are nice people.**"

Sure you are, but sorry, I'm just too busy. Using another of my noms de plume, somewhere@around.com, I'm apparently participating in this week's Farmers Market Online Poll. Someone in London is trying to persuade me, thardy@around.com, to buy "**the gizmo that everyone's talking about,**" namely, the Executroid Relaxation Module ("**Stay tuned for more info!**"). I'm browsing new cars and real estate. Janet Chong in Singapore has rushed over my year 2000 horoscope, but one of the side effects of my multiple-personality disorder is that I've lost track of my sign. As yosh@around.com I am accumulating "Points" at a place called Talk City; I fear these points are as imaginary as I am.

Then, as Bozo the Clown (clown@around.com), I have submitted a service request, "**Priority level top**," from the Blanco School's North Room Gym in the Port Orford–Langlois school district in southern Oregon. How old am I, anyway?

Anonymity had a long and happy run in pre-Internet times. A free society cherishes whistle-blowers and dissidents. We know we aren't all fortunate enough to live in communities that protect unpopular forms of expression. We know that anonymity can free us to seek help for drug abuse or mental illness. We may be slow in recognizing, though, how profoundly the character of communication is changing—and quite suddenly. A fine set of laws and rules of etiquette evolved over centuries to cover modes of expression that we now see were slow and intimate and short-range: voices in the town square, notices on the church door, petitions, letters, newspapers.

Now these rules must cover cyberspace, an instantaneous global democracy, where everyone with a keyboard has a voice, and every voice carries far, and every word lingers in archives and databases with startling persistence. Many semipublic commissions and policy institutes have been analyzing the issues raised by online anonymity since the first days of the Internet; the consensus always ends up sounding like a high-toned good news, bad news joke. Rob Kling, professor of information science at Indiana University, introduced a special issue of the journal *Information Society* this summer with the canonical version of anonymity's "double-edged character": on the one hand, a fanatic signing e-mail as Asian Hater spreads fear at the University of California by threatening to hunt down and kill fellow students; on the other hand, human rights activists use anonymizer.com to protect Kosovars trying to send out reports on the Balkan war.

We understand the dilemma. On the one hand, freedom from persecution and embarrassment; on the other hand, a mask for criminal and antisocial behavior. On the one hand, the Lone Ranger; on the other hand, the Ku Klux Klan. On the one hand, hoaxes, libel, and fraud. On the other hand . . . free speech?

But free speech never used to mean *nameless* speech. It has not, until now, entailed the power to broadcast anonymously to millions of listeners. Creating protection for unpopular expression was important, and expensive, precisely because everyone knew who those unpopular speakers were. In the ancient small world we know from history books and black-and-white movies—when people stayed in more or less the same place and encountered pretty much the same dramatis personae year after year—a person's name mattered. If someone's character was tarred or besmirched, the damage wasn't easily undone. Reputation stayed with you.

Then we noticed that our big, crowded, urban world was different—anonymous. Communication by telephone is faceless. We learn early that it's safe to phone the store and say, "Do you have Dr Pepper in a can? Let him out!" At first the Internet looks like more of the same—more crowded, more wired, faster, just as faceless. It turns out that cyberspace works differently. Communities form, here and there, freed from constraints of geography;

and that old small-town feeling emerges. In the most successful communities, identity and reputation carry significant weight. The auctions at eBay depend, as a matter of trust, on the site's elaborate system of recording comments by members about other members. Elsewhere the aficionados illegally trading digitized jazz recordings get to know one another; as do the nervous obsessives arguing about the year 2000; as do the stock traders spreading rumors on Yahoo! about California chipmakers and Florida banks and just about every other public company. But can they believe what they think they know?

It turns out that half the recent messages about Harbor Florida Bancshares came from the keyboard of one Maytricks99— **"storm the Bastille, er, the downtown Fort Pierce office, and demand your $42 per share"**— and management has filed suit against him, naming him, quaintly, as "John Doe." Xircom, a modem maker in California, won a similar suit in June over pseudonymous attacks by a critic claiming to be a company engineer; Xircom's John Doe appealed, and then settled with the company, admitting that he had never worked there. Many such suits are under way. However the courts end up resolving them, the participants in company-oriented message boards at Yahoo! and elsewhere are the losers, because they will quickly learn that they can't take this form of communication seriously. The people touting and savaging companies are mostly owners and sellers of the stocks, lobbing their little grenades online in hopes of influencing the price.

Worse than that, the growing use of false identity as a standard mode of online behavior suggests a sort of self-loathing, on a mass scale. It seems we so hate the images we project that we feel compelled to detach our names from them. We know we're being mean-spirited, offensive, irrelevant, or noisome with our con-

freres in the TimeBomb 2000 discussion forum, so we cloak ourselves as "Bulldog" and "Z1X4Y" and "Sniffin" and "Peon." We're
not acting proud, anyway. Also, to stay truly anonymous online
is harder than most people think.

It's easy to cook up a false identity, even an elaborate one, in a
matter of minutes. Giant free mail services like Hotmail give out
usable addresses for the asking. They require your name, but go
ahead, tell them you're Boris Badenov—they don't mind. They
seem to be daring you. Netscape's Netcenter has actually touted
its mail service with the slogan, "Send a nasty note to your boss."

Nevertheless, people leave electronic footprints. Service
providers usually have our credit card information. Web browsers
give away more about your PC and your Internet location than
you might wish. Microsoft, which owns Hotmail, wants you to
know that it's not meant as an anonymous e-mail service. The
service keeps its records associating users with their computers'
Internet addresses. "You're traceable," says Laura Norman, a
Microsoft product manager. "You can be traced, and if you do
something that's illegal or harmful we have that history and *will*
turn it over to law enforcement." Not everyone is sophisticated
enough to know this. Not even everyone at Microsoft. Just a few
days after Norman told me this, another Microsoft employee got
caught using a Yahoo! mail account to masquerade as one Phil
Bucking of Bucking Consulting, a fictional concerned citizen
sounding a fictional alarm about security issues at America
Online.

Various companies are offering various high-tech solutions
to this problem—the problem of imperfect anonymity. Zero-
Knowledge Systems, for example, markets a product called,
simply enough, Freedom, hoping to "redefine identity on the
Internet" and "create authenticated digital pseudonyms that bear

no necessary relation to the actual person behind them." It claims to use cryptography to cover your trail as you manage all your different identities. "For example," the company suggests, "if you like to debate politics online you can designate one pseudonym as your 'politics' pseudonym. Use it when you post in political newsgroups, surf activist Web sites, e-mail your political contacts and chat in political chat rooms. No one can trace it back to your real self."

If you still have a real self. Maybe you have a new multiple-personality disorder instead. A meager and unpleasant discourse emerges from the jousting of all these pseudonymous role-players.

Hardly anyone thinks that governments should try to regulate matters of identification online. Nor is it likely that governments could, even if they wished to. Still, it's worth remembering that the Internet is young. Much of its structure exists as pure accident. Widespread anonymity was not part of the plan. When the basic protocols of internetworking were laid down, openness prevailed; you could "finger" people electronically and see not only their names but whether they were online and when they had last checked their mail. It would be entirely possible, and eminently reasonable, for individual Internet service providers (private companies, after all) to require their credit card—carrying customers to use real names for their public activity. It's technically feasible for e-mail servers to stamp outgoing messages with a valid return address. The various grotesque mutant species of electronic junk mail would probably wither away if this became a universal habit; for now, the senders use anonymity as a tool for invading your privacy.

Too many masks. Maybe we can't really run from our names, in this peculiar modern realm. Or maybe we just won't like the neighborhood if we try. A new message arrives for one of my fictive selves. "**We have received a request to put this email account on hell.com's guest list**," it advises.

No thanks. I'll pass.

It's time to close all those mailboxes.

PATENTLY ABSURD

March 2000

When the twenty-first century reflects on the breakdown of the United States patent system, it will see a turning point in the case of Jeff Bezos and Amazon.com and their special *invention*: "the patented 1-Click feature," Bezos calls it.

Not everyone who knows Bezos as the new-minted billionaire founder of the world's leading Internet retailer knows that he's also an inventor, but he is. It says so on U.S. Patent No. 5,960,411, "Method and System for Placing a Purchase Order via a Communications Network." Every good invention needs a story, and Jeff Bezos has one for one-click ordering. He's laying it out in federal court, where already, at the height of the holiday shopping season, he won an injunction forcing his chief competitor, Barnesandnoble.com, to add deliberate complication to its ordering process.

In ways that could not have been predicted even a few years ago, the patent system is in crisis. A series of unplanned mutations have transformed patents into a positive threat to the digital economy. The patent office has grown entangled in philosophical confusion of its own making; it has become a ferocious generator of litigation; and many technologists believe that it has begun to choke the very innovation it was meant to nourish.

The one-click story starts like this: A young man founds a company called Amazon.com in the obligatory garage, with a bit of money raised from his family. Soon his tiny crew begins selling books online. Their customers use a "shopping cart." Of course, there isn't a real shopping cart, because the customers aren't really *at* these stores, and there aren't really *stores* at all. There is just a tiny picture of a shopping cart, and if you click on it with the mouse, the screen displays a list of items you have selected for purchase.

But the online shopper is thought to possess a thoroughly modern attention span, and many shopping carts are left abandoned in their virtual aisles. So Bezos decides ("sometime prior to May of 1997," he says with the help of his lawyers) to offer a shortcut: let users choose an item with just one click of the mouse button, and use a shipping address and credit card number already on file. Obvious? Sure, but among his employees some are naysayers, worried about spooking the customers. "Conventional wisdom was that they had to be slowly and incrementally led to the point of purchase," Bezos recalls. But his vision will not be denied. He orders his software developers to "make it happen." And so they do.

Amazon applies to the government for a trademark on the name "1-Click" and a patent on the . . . well, on what, exactly? Not the idea, because pure abstractions may not be patented. Not the program code, because copyright law protects this. But Amazon received its patent in September and instantly sued Barnes & Noble over its similar "Express Lane." Amazon won an injunction in December forcing its competitor to insert a superfluous mouse click. ("Please be sure to click this button," Barnes & Noble begs plaintively. "If you don't, we won't get your order!")

The one-click injunction capped a burst of skirmishing in 1999—the start of what promises to become furious, wide-ranging courtroom warfare—over who will control electronic commerce. As the year 2000 begins, few of America's e-commerce leaders are *not* targets of patent litigation. The battles to come will determine whether the essential tools and building blocks will continue to spread rapidly through the community of software designers and Internet pioneers, or whether they will be cordoned off as the private property of particular companies. Amazon just got a new patent for a system of letting Web sites refer customers in exchange for commissions—its "affiliate" program. Sure enough, Barnes & Noble has a nearly identical affiliate program. So do thousands of other online merchants, now feverishly calling their lawyers yet again. Can Amazon really *own* this? For better or worse, the struggle will redefine our understanding of what an invention is, in our complex, technocratic age.

Patents long served as a fundamental cog in the American machine, cherished in our national soul. We are the land of Thomas Edison and the Wright Brothers and Alexander Graham Bell, where Congress is empowered by the Constitution *to promote the Progress of Science and useful Arts, by securing for limited Times to Authors and Inventors the exclusive Right to their respective Writings and Discoveries.* Hence the patent office, charged with the enforcement of a Faustian bargain: inventors give up their secrets, publishing them for all to see and absorb, and in exchange they get twenty-year government-sanctioned monopolies on their technologies. This arrangement fueled industrial progress in the early United States by encouraging investment in research and rewarding inventors who published their work rather than cloaking it in trade secrets.

Now, however, during the short span of the Internet revolution, the patent system has begun to disintegrate by growing out of control. The United States is issuing patents at a torrential pace, establishing new records each year, and it is expanding the universe of things that can be patented. Patents began in a world of machines and chemical processes—a substantial, tangible, nuts-and-bolts world—but now they have spread across a crucial boundary, into the realm of thought and abstraction. Software and algorithms used to be unpatentable. Recent court decisions and patent office rule-making has made software the fastest-growing patent category, and companies are rushing to patent the most basic methods of doing business. "This is a disaster," says Lawrence Lessig, a Harvard law professor and cyberspace expert. "This is a major change that occurred without anybody thinking through the consequences. In my view, it is the single greatest threat to innovation in cyberspace, and I'm extremely skeptical that anybody's going to get it in time."

The litigation is spreading fast. Multi-Tech Systems has just sued the three leading PC makers, Compaq, Dell, and Gateway, over patents on transmitting data over a communications line. A St. Louis patent broker is suing Yahoo! over a "method of effecting commerce in a networked computer environment in a computerized system"—that is, shopping online. Another Internet start-up, Priceline.com, has patented its Internet version of an ancient auction technique, the name-your-price "reverse" auction, and is suing Microsoft's Expedia.com travel service. Microsoft, meanwhile, has infuriated much of the Internet community by patenting a well-known "style sheet" technology just as it was being adopted as standard by the World Wide Web consortium.

Sightsound.com is suing at least one music retailer and demanding royalties from others over a patent on selling audio and video recordings online. A California software company is suing eBay over database technology. Two professors at the Massachusetts Institute of Technology are suing the Ask Jeeves search site over two patents on handling questions in natural language. A Boeing software engineer has patented a basic method of correcting the century in dates stored in databases and sent a threatening form letter to seven hundred of the nation's largest corporations, demanding one-fourth of a percent of their total revenues, on the assumption that they probably used the same method.

This is just the beginning. Patents marking off broad swaths of electronic commerce will soon be pouring from the patent office, unwelcome surprises to whole categories of new entrepreneurs. In the last few months, companies have gotten patents for keeping calendars on the World Wide Web, for downloading Web pages at regular intervals, for storing documents in databases, for "real-time shopping," for auctioning cars, for creating profiles of users; for search engines, for payment systems, and for variations of every other fundamental gear and lever in the theoretical machinery of online business. Just as insidiously, biotech companies are getting thousands of patents granting them rights to exploit particular pieces of the human genome—the DNA common to all of us.

For that matter, the most trivial slices of off-line life are winning patent protection: for example, "measuring breasts with a tape to determine bra size"; and "executing a tennis stroke while wearing a knee pad" (U.S. 5,993,336: "The tennis racket is swung toward a tennis ball so as to hit the tennis ball with the racket . . .") Many of these patents are harmless. Most are narrower, when read carefully, than they sound at first. Others are multimillion-dollar lawsuits in embryo. Every week, hundreds of new data processing patents are issued, and if you are an entrepreneur with an e-commerce business plan, you will soon be cross-checking it against these patents. Won't you be using a "metering mechanism for distribution of electronic information"? (Intel owns it.) "Tracking the purchase of a product and services over the Internet"? (InfoSpace.com.) A "method and system for constructing queries"? (Microsoft.) "Expanding Web documents by merging with linked documents"? (IBM.)

Each of these patents is a tiny masterpiece of logic and disputation. Each represents, by definition, a restraint on trade, a layer of regulation, expensive overhead in the free-market economy. The exclusive rights conveyed by a patent automatically translate into higher prices for consumers somewhere along the chain. The Supreme Court saw the downside more than a hundred years ago. "It creates a class of speculative schemers who make it their business to watch the advancing wave of improvement, and gather its foam in the form of patented monopolies, which enable them to lay a heavy tax upon the industry of the country," the Court wrote. "It embarrasses the honest pursuit of business." Patent lawyers and officials argue that each patent represents a good idea and that good ideas deserve to be rewarded; but does every good idea deserve a twenty-year government-sponsored monopoly?

"We like to say 'right to exploit,'" Commissioner of Patents and Trademarks Q. Todd Dickinson says cheerfully.

• • •

Patents are widely supposed to protect the lone inventor, the pioneering genius in a garage, against the predation of big companies. Historically the opposite has been true. As basic industries like electricity, telephony, and broadcasting developed in the twentieth century, the great corporations learned to create arsenals of interrelated patents to use as sword and shield. "The wise people, with good patent lawyers, patent a whole system," says Thomas P. Hughes, a historian of technology at the University of Pennsylvania. Although small companies can get patents, the big companies can afford to litigate—"when it comes into court, guess who's going to win." Absurd patents can be fought and overthrown, but, on average, to challenge a patent costs more than $1 million.

"Even under traditional patent rules, many of these software patents will turn out to be bad patents," says Lessig at Harvard, "but in the meantime they create these little mafia monopoly holders who can go around demanding, with a federal court behind them, that you pay up or we'll shut you down."

Patent battles have become a strong catalyst for mergers, reducing competition in various domains. The largest corporations, with gigantic patent portfolios, routinely enter into cross-licensing agreements with their largest competitors. Companies without portfolios of their own have to pay cash, representing a sort of tax within the high-tech economy, and the numbers are skyrocketing: costs in the United States for patent licenses were about $15 billion in 1990; eight years later they had soared to more than $100 billion. IBM alone took in well over $1 billion last year and received a record 2,756 new patents. The top ten recipients of U.S. patents also comprised Motorola, Lucent (home of the former Bell Laboratories), and seven Asian computer companies.

Those lone inventors are out there nonetheless, dreaming of

the patent that will make them rich, and they are a familiar species of fraud victim. An entire industry of invention promoters promises to help inventors get patents, usually charging thousands of dollars in fees that are virtually never recouped. Hopeful inventors show up in person at the patent office in Crystal City, Virginia, many returning month after month, some eschewing the all-day application window to bring their files directly to the commissioner's office.

Inventors rank high in our pantheon of cultural heroes. We tend to think they deserve rich rewards: if someone has a great idea that makes life a little better for millions of people, surely fortune and fame are fair compensation. Yet, when it comes to rewarding and protecting the greatest achievements, the history of twentieth-century invention suggests that the patent system has at best a mixed record. A few milestones:

The transistor. Probably the single most important invention of the century. William Shockley, John Bardeen, and Walter Brattain invented it at the Bell Telephone Laboratories in New Jersey, and they received patent No. 2,524,035 for it: "Three-Electrode Circuit Element Utilizing Semiconductive Materials." They also won the 1956 Nobel Prize in physics—considerably more useful, from the fame-and-fortune point of view, because Bell Telephone already had enough of a monopoly, and the world has gone on to manufacture billions of transistors daily without paying licensing fees. It would be hard to argue that that's either unfair or bad for the global economy.

The communications satellite. A great idea traceable to a single inventor, Arthur C. Clarke, a twenty-eight-year-old Royal Air Force officer later to become famous as a science fiction writer. He sketched the whole idea, with detailed calculations and drawings, in the journal *Wireless World* in 1945; the small article fee was all he ever got. In his slow-paced world, the timing was wrong for a patent—the idea was years from becoming practicable. "It is with somewhat mixed feelings," Clarke wrote later, "that I claim to have originated one of the most commercially valuable ideas of the twentieth century, and to have sold it for just £12."

Velcro. Occasionally the system does work as we imagine. Georges de Mestral, a Swiss inventor, conceived of the now-indispensable hook-and-loop fastening technology in 1948, not in his garage, but supposedly after getting his pants entangled in cockleburrs and examining them under a microscope. He patented a version of what he saw and founded a company that made him extremely rich.

The World Wide Web. Tim Berners-Lee invented the Web and the Web browser—that is, the world as we now know it—pretty much single-handedly, starting in about 1989, when he was working as a software engineer at CERN, the particle-physics laboratory in Geneva. He didn't patent it, or any part of it. On the contrary, he has labored tirelessly to keep cyberspace open and nonproprietary.

So this is the Patent That Never Was—its *non*existence directly responsible for the growth of cyberspace. "Anyone who puts a small gloss on this fundamental technology, calls it proprietary, and then tries to keep others from building further on it, is a thief," says the Internet pioneer Tim O'Reilly. "The gift was given to all of us, and anyone who tries to make it their own is stealing our patrimony."

•　　•　　•

At first, the awarding of a patent was special—a profound and unusual act. Under the original patent law of 1790, the examiners who met to consider each invention were none other than the secretary of state, the secretary of war, and the attorney general: thus Thomas Jefferson, Henry Knox, and Edmund Jennings Randolph awarded the first U.S. patent on July 31 for a process of making potash. They issued two more patents that year. Gradually the patent system grew and became associated with the American inventive spirit. Patents seemed not only to reflect the young nation's technological genius but to have fomented it. At a time when information moved slowly, patents created an unparalleled storehouse of how-to data about technology, and a pipeline for sharing that information.

They also became a source of prestige and distinction; after all, to be an inventor was to be a patent-holder, and vice versa. "The patent system added the fuel of interest to the fire of genius," said Abraham Lincoln, himself the proud holder of U.S. Patent 6,469 ("Device for Buoying Vessels over Shoals"). Even so, it took forty-six years of growing bureaucracy and accelerating output to produce a total of 10,000 American patents. Now the patent office issues that many patents every three weeks. A single examiner can approve a patent, without review from supervisors or the commissioner. "If an examiner allows a case," one examiner says, "not even the president of the United States can force him to change his mind. Then there's the Supreme Court, and they're kind of busy." Each examiner is a specialist, meant to be familiar with the landscape and the "prior art" background for particular types of technology. But for the ethereal new realms of software and business methods, especially, the system has broken down. Through most of the young history of software engineering, the state of the art has been carried around in the heads of young programmers working late nights in offices strewn with soda cans and pizza boxes, not in academic journals suitable for indexing and perusal by patent examiners. Examiners give most weight to their own database, anyway, treating the 6 million existing U.S. patents as a sort of filing cabinet of all human knowledge.

"The U.S. patent office is just not competent to examine software patents," says Gregory Aharonian, a consultant and publisher of a widely read patent newsletter. "Eighty percent of software patents effectively cite nothing from the computing literature. To me it's a kind of contempt." He contends that the patent office has neither the time nor the expertise needed to distinguish good patents from bad.

"It's a cold war," he says. "It's just people playing legal games."

The patent office has become a place where the essential politics—the checks from contending interests—are out of balance. The voices heard daily at the patent office belong to people who like patents, want patents, and rely on patents for their living; their creed is, *the more the better.* Officials measure their own performance in terms of their output. It's as if they were a manufacturing company turning out product.

The agency is proud, too, that foreign companies have been stepping up their applications for U.S. patents on their software and business-method ideas; so far, the European and Japanese patent systems have been less willing to grant protection for these. Meanwhile, the dollars-and-cents reality of running the American patent office has also encouraged the patent explosion. In 1991, the patent office was cut off from general tax revenues and required to subsist entirely on fees for its operating budget. The political argument was that customers should pay for government services. Thus, officials think of their fee-paying patent applicants as their customers: *the more the better*, again. Examiners know that their year-end bonuses depend on productivity. The people interacting regularly with patent officials and examiners—their obvious clientele and customer base—are inventors and inventor representatives.

Each morning, as Commissioner Dickinson arrives at his Crystal City office, he walks past a framed poster bearing the motto:

Our Patent Mission
To Help Our Customers Get Patents

It's virtually forgotten that the government's *customers* also include the rest of the nation, the citizenry at large, whose fortunes depend on the agency's judgments and policies.

With the advent of computing, human invention crossed a threshold into a world different from everything that came before. The computer is the universal machine, almost by definition, machine-of-all-trades, capable of accomplishing or simulating just about any task that can be logically defined. Even more so with cyberspace—everybody's computers, connected—the universal machine to the *n*th degree. Take any real-world procedure or activity; transfer it to cyberspace; and now we have an activity that seems both new and technological. Patents have recently been awarded for selling airplane tickets online; selling software online; "network sales systems" (many); "distributing advertising over a computer network," and countless more.

But patents played no role in the early history of personal

computer software. Microsoft, having been founded in 1975, received its first patent in 1986—for a sort of hinged box. Another two years passed before it got its first computer-related patent. IBM and some other companies were patenting software, but they had to engage in a careful sophistry to do it: making sure to say that a program was an *apparatus* or a *system* and that it was *embodied on a computer-readable medium.* Applicants and examiners employed a mutually understood pretense that machinery was still involved.

The biggest battles over intellectual property in the software business focused on copyright. Apple sued Microsoft in 1988 for copying the "desktop metaphor" and other key elements of Windows, and Lotus sued Borland in 1990 for copying menus and command sequences in its spreadsheet software. Both defendants had clearly copied something, but it was not the actual code; it was something vague, in the nature of "look and feel," and ultimately the courts ruled in favor of the defendants. For books, copyright law protects actual text, not ideas or plot; the law had to figure out what that meant for this new form of expression, software. Some software authors in the 1980s tried to extend their copyright protection beyond the literal code, to elements at higher levels of abstraction: a common phrase was "structure, sequence and organization." Ultimately the courts held that copyright law did not cover these broad elements. The computer industry itself was divided; so was the Supreme Court, which split 4–4 in refusing to overturn the *Lotus v. Borland* ruling.

Patent officials like to suggest now that those cases might have ended differently if Apple and Lotus had patented their innovations. The decisions "tilted the system away from copyright," Dickinson says. "People shifted and started to use the patent system."

As the software industry began flooding the office with applications, the system came around to a view of software programs as machines. They *are* machines, in a way. "Anything that can be done in hardware can be done in software," says one examiner. "We're not using gears; we're using some other mechanism, some other means." The patent office used to require an exact model of every invention it considered. Edison's original light bulb is sitting in Commissioner Dickinson's office, near a window with a fine view of Reagan National Airport. The models accumulated until the government ran out of storage space and abandoned the requirement. Lucky thing, because these new machines are machines without substance—incorporeal machines, machines made of imagination and logic. No one will be building models of these airy phantasms of bits.

One more crucial court decision came in July 1998: the State Street Bank appeal. State Street had been sued for infringing a 1993 patent for an intricate computerized strategy of managing a multitiered portfolio of mutual funds. A federal court in Massachusetts found that the real subject of the patent was not a "machine" but a business method in data processing garb, and declared the patent invalid. Mathematicians and physicists sometimes argue about whether they *invent* or *discover* the laws of nature. Did Einstein invent his formula $E = mc^2$ or was it there all along, waiting to be discovered? He couldn't have patented it, the Supreme Court has stated: such laws are "manifestations of nature, free to all men and reserved exclusively to none." Judge Patti B. Saris cited that principle in the lower-court decision, noting some other key precedents: "Mental processes, and abstract intellectual concepts are not patentable, as they are the basic tools of scientific and technological work"; and "an improved method of calculation, even when tied to a specific end use, is un-patentable subject matter."

Saris decided that the so-called invention was no more than a way of calculating: "The same functions could be performed, albeit less efficiently, by an accountant armed with pencil, paper, calculator, and a filing system." The patent gives its owner a monopoly, she pointed out: "patenting an accounting system necessary to carry on a certain type of business is tantamount to a patent on the business itself."

This was thoughtful, and it was right, but the court of appeals foolishly reversed that ruling in 1998 and reinstated the patent. Even "abstract ideas constituting disembodied concepts or truths" could be patented, it declared, as long as they performed a useful function. The useful function in this case was to produce "a final share price." In other words, the court ruled, software that merely manipulates numbers, juggling them and exchanging them and transforming them into other numbers, is producing something *tangible*. A powerful conclusion, in a digital age. It opened the floodgates.

There is a sense of infinite regress to the argument, a feeling of abstraction upon abstraction, words like *algorithm* and *formula* and *process* and *method* defined and redefined circuitously, if not circularly. Judges and lawyers have devoted millions of words to the nuances. Maybe the trajectories of culture, economics, and technology have reached a point where a distinction between *idea* and *machine* can no longer be sustained; where no bulwark of logic, but only the mist of undecidability, separates $E = mc^2$ from the light bulb.

•   •   •

By the time of the State Street decision, Amazon's application had been working its way through the patent office for nearly a year. Barnes & Noble, late to the Internet business and lagging far behind Amazon, had redesigned its site yet again. Its new Express Lane looked to Bezos like a clone of his one-click system. Indeed, anyone examining the faces these sites present to the world might conclude that Barnes & Noble has copied a lot more than just one-click ordering. Both sites let customers post reviews of books; both display an average of the customer ratings in the form of one to five stars. They both list books that "customers who bought this book also bought." In October 1998 Amazon filed a "petition to make special," asserting that an "infringing product" was already on the market. Barnes & Noble didn't know that at the time; pending patent applications are kept secret.

In Crystal City, the examiner, Demetra Smith, was studying a small selection of prior art—a few earlier patents and articles from the computer trade press—trying to make sure she understood the difference between Amazon's invention and earlier shopping-cart systems. The key to the one-click system is actually quite simple. If a customer has shopped at Amazon before, credit card and other information is stored in the company's computers, and meanwhile the customer's Web browser has stored a special file called a *cookie*. That's how the browser knows, when the customer returns to Amazon's site, to display a personalized message: "Hello, Todd Dickinson! (If you're not Todd Dickinson, click here)." Cookies were already widespread in e-commerce, and the examiner had among her papers a 1996 description of how Netscape's software implemented them.

As far as the examiner could see in the papers before her, no one had talked about using cookies to let customers skip past the whole shopping-cart checkout process, so in a sense Amazon's idea was, in fact, new. Still, she could see the basic elements in a patent that Lucent Technologies had filed just a few months before Amazon, describing "user identifiers" sent back and forth on a network. She noticed a comment that the identifiers help a company "to recognize a returning user and, possibly, provide personalized service." Just what Amazon does! The next step seemed *obvious*—in the patent office's technical sense of the word. "It would have been obvious to one having ordinary skill in the art at the time the invention was made," she ruled tentatively, "to include various command mechanisms for a single user action in order to execute the user's request."

Obviousness is the key problem. "How do you tell, in a new field of the economy, what is obvious?" says Joseph Farrell, an economist at the University of California at Berkeley. "If you have a wide-open field suddenly opening up out of nowhere, there may be a lot of things just waiting to be done which are pretty obvious, and there's no need to make them patentable to provide incentives for people to do them. And then it's kind of a pity if they get patented."

Every patent examiner's work is about making fine distinctions. Is that "ergonomic topographic toothbrush" novel enough to merit a new patent of its own? These are delicate enough questions when the subject is real stuff. Patent examiners can sink their teeth into concrete details like a "detachable and replaceable bristle head." With software, where the nuts and bolts are vaporous and intangible, questions of what's *obvious* and what's *novel* begin to float in the wind. In any event, the burden of proof is on the examiner to show that an application must be rejected. Her subjective judgment is not allowed—only actual references in published literature. "People send in some really strange stuff for patents, and I have no choice but to issue it," says one examiner. "I can't say, gee, that's obvious to me." Evidence of obviousness "has to be out there and public and in the same detail."

So Amazon's lawyers argued back, pointing out that the Lucent patent specified anonymous browsing and nowhere mentioned single-action ordering. They negotiated, agreeing to drop some claims in their application and amend others, and they followed the examiner's directions for reworking their drawings (margins incorrect, numbers too small, lines too irregular), and in the end, like most patent applications, this one was approved. After all, examiners are motivated to issue patents, not to block them. U.S. 5,960,411 grants Amazon exclusive rights to its method of placing an order "in response to only a single action." Also, ". . . wherein the single action is clicking a button." And "the single action is speaking of a sound." And wherein a user "does not need to explicitly identify themselves." And "the single action is selection using a television remote control" or "a pointing device" or "depressing a key on a key pad." Amazon has itself pretty well covered. Then again, hundreds of new e-commerce patents will be issued in the coming month, just about all to Amazon's largest and toughest potential competitors. Until they are issued, neither Amazon nor any other company will know whether it is infringing them—perhaps with an innocent and obvious bit of homegrown programming.

Amazon claims that it spent thousands of programmer-hours on its one-click method, but the patent system doesn't care about that. In determining what may and may not be patented, the law does not distinguish between inventions that require expensive research and inventions that amount to a momentary flash of insight. A new drug that costs ten years and millions of dollars of research gets the same protection as a bit of programming that comes to a lone hacker in a dream. In purely economic terms, this is inequitable and perhaps even dangerous.

"We're not talking about Thomas Edison inventing the light bulb," says Lessig. "We're not talking about Monsanto spending tons of money on some chemical whatever. We're talking about people taking ways of doing business and, because they put it into software, they say, 'This is now mine.'"

Amazon won't say how many patents it has pending. The one-click patent isn't its first, as it happens; Jeff Bezos got one in February 1998 for "a method and system for securely indicating to a customer one or more credit card numbers that a merchant has on file for the customer when communicating with the customer over a non-secure network." The method is this: show the customer only the last few digits of each credit card number.

Many companies use this very method, of course. Even if they thought it up all by themselves, they may be infringing Bezos's patent. He and Amazon won't say whether they've begun trying to collect licensing revenue.

All around, hopeful entrepreneurs are treasuring new patents like geese ready to lay golden eggs. Stephen Messer, for example, is building an Internet business called LinkShare around a patent he got in November on managing referral fees for traffic between affiliated Web sites. His sister, Heidi Messer, a lawyer who is now LinkShare's president, encouraged him to apply for the patent on his idea. "When he told it to me, I thought it was profound," she says.

He's a believer in patents now. "Without the patent system, I never would have done what I did," he says. "Without the patent system you create a world where pirates have the advantage and pioneers are penalized." His lawyers, in fact, are taking a close look at the many companies using affiliate-referral systems. One of them is Amazon.

As for one-click ordering, a final verdict could be years away, and in the meantime, online merchants will have to be careful. Mothernature.com already has one-click remedy packs. Safe Harbor has one-click ordering of digital video products. Are they patent infringers?

"I think the beauty of the patent law is, it encourages innovation, because it says, look, we challenge you to come up with a better way," says Heidi Messer.

"Somebody out there," says Stephen, "I guarantee you, probably sitting in a garage right now, has an idea that does zero-click."

Heidi laughs. "Telekinetic!"

Telekinesis would be worth patenting. Meanwhile, much of the value of the patent system lies in the disclosure of technologies that might otherwise be hoarded as trade secrets. But when Barnes & Noble decided to implement one-click ordering, its programmers did not need to see Amazon's patent. The software techniques are transparent. Any Internet company with a decent programmer at hand could imitate Amazon's system without breaking a sweat and without copying its actual code. This isn't rocket science, in other words. I say as much to Commissioner Dickinson. He is unruffled; evidently he's heard it before. "We don't patent only rocket science," he says. "We patent football-helmet mailboxes." And indeed they do. Sportsbox Inc., aka a couple of guys in Richmond, Kentucky, owns the patent for mailboxes in the shape of football helmets; before you make one, you'd better get their permission.

Dickinson, who practiced as a corporate intellectual-property lawyer for twenty-five years before taking over the patent office, says the law prohibits him from discussing any particular patent. (The patent office forbids its examiners to speak to the press at all, and the examiners who did talk to me about their work insisted that their names be withheld.) Still, he can't resist revealing some pride in the progress of the Amazon case so far. "It's interesting because it's one of the first Internet patents to come to at least one conclusory stage of the litigation process," he says. "A federal judge found that the patent was sufficiently valid to issue a preliminary injunction, which is a rather extraordinary thing, right in the face of the Christmas holiday season. I thought that was kind of amazing, to be honest, and tends to suggest that the patent has validity."

Barnes & Noble tried to assemble earlier references to elements of the one-click method from the scattershot literature of the early dark days of e-commerce, 1995 and 1996. They found descriptions of shopping carts and web baskets and virtual stores, but nothing that looked to the judge, Marsha J. Peckman in Seattle, like a single-action system. There was an old CompuServe system of selling stock-price quotations, but it was not an Internet service; users had persistent connections to CompuServe, so there was no need to use cookies to identify them. Besides, they had to type in ticker symbols. "The Court finds that this method involves two actions, not one," Judge Peckman ruled solemnly.

In making her preliminary finding that Barnes & Noble infringed the Amazon patent, she did not have to examine the underlying software code. Just *how* Barnes & Noble implemented one-click ordering doesn't matter. Any single-action ordering system violates Amazon's exclusive rights, she found. She also expressed a public-interest argument: "Innovation will be discouraged if competitors are permitted a free ride on each other's patented inventions," she declared.

The commissioner agrees: "The patent system has done its job for two centuries of protecting and nurturing and rewarding innovation. The system has worked."

But the digital revolution worked *without* patents. The great bursts of technological innovation of the past two decades, the rise of personal computer software and the spread of the Internet, took place in a freewheeling and competitive climate, with ideas bouncing at light speed from one place to another. A little head start in this world goes a long way; in the digital economy, "first movers" gain a tremendous and possibly long-lasting advantage, without extra government fortification. The greatest successes, like Microsoft and America Online, had nothing to do with patent protection. Amazon did not need patents to grow from Jeff Bezos's garage to its current preeminence.

Only the Congress can now return the patent system to its time-honored role as a catalyst for innovation, rather than a concentrator of economic power. *For limited times,* the Constitution says—and the generations of technology pass much more quickly now than they did in 1790. We want to reward inventors with a head start, not a lifetime entitlement. "Perhaps on the Internet, patents should last one Internet lifetime, which is about two years," says Greg E. Blonder, longtime researcher and vice president at Bell Labs and now entrepreneur in residence at AT&T Ventures.

This is a crowded and densely packed world; when a software engineer at IBM comes up with a new idea, it's usually safe to bet that dozens or hundreds of software engineers in cubbyholes around the world are thinking along the same lines. Blonder himself has more than sixty patents to his name: toys, consumer electronics, software, and business methods. He often gathers technologists for daylong brainstorming sessions; typically, these groups produce hundreds of useful "ideas," of which perhaps one in ten turns out to be truly novel. "Problem is," Blonder says, "had I gathered together a different group of a dozen people, they'd come up with practically the same list. Arguably, any idea generated so easily and frequently is both obvious and a dubious candidate for a twenty-year government-sanctioned monopoly." Patents should be the exception, not the rule, he says. "As to my own business-process patents, well, as long as everyone in town is carrying a gun, I have to be armed as well. But I'd be glad to see the system change."

One thing is certain: the modern Internet entrepreneur is not a species in need of extra government incentives. If one-click ordering had not been patentable, surely Jeff Bezos would have invented it anyway in May 1997, put it to work in September 1997, seen it copied in subsequent years by Barnes & Noble and thousands of other Internet merchants, learning from his success, to the overall benefit of consumers. Surely he would have become a very rich man; and his future success would depend on his ability to continue outperforming his competitors, and to continue innovating.

INESCAPABLY CONNECTED

April 2001

As I drive my rental car across Silicon Valley under a cloudless and starry sky, it is fitting that the electronic navigation device on the dashboard should be talking to me. "*Approaching left turn,*" says Helga (as I call her). "*Left turn in point five miles.*" Headlights rush past us, exit signs loom and are gone, and now it occurs to me that this freeway doesn't even have left turns. Helga is trying to show me something on a tiny, color-coded, icon-studded moving-map display at the edge of my peripheral vision. Up in the real world, we hurtle under an overpass. *That* wasn't my left turn, I hope. But yes, apparently Helga lacks perfect knowledge of California cloverleaf topography. "*Calculating route,*" she chirps, as if we can simply begin again, with no memory of the past. I am mindful of the German motorist who drove his BMW straight into the Havel River one night because he put too much trust in his dashboard navigator.

Still, we can't get lost. We're too well connected, Helga and I. She listens constantly to at least four of the two dozen satellites of the Global Positioning System: orbiting atomic clocks that bathe the globe in their precisely intermingled time signals, enabling any device skilled in trigonometry (and these days what device isn't?) to reckon its exact location. We're not alone here. My cellular phone, as long as it's on, parleys silently with the network, giving and receiving information about when and where we are. My handheld Palm-type computer cum wireless modem has already pulled in directions by e-mail and can download new maps in real time. I could plug my laptop computer into the cell phone, or vice versa, and be online that way. (I haven't felt the need to give all these devices names; most of them aren't talking to me.)

The network knows where we are. The network is there, all around us, a ghostly electromagnetic presence, pervasive and salient, a global infrastructure taking shape many times faster than the interstate highway system or the world's railroads. This is different from the radio-spectrum Babel that defined the twentieth century: the broadcast era. We aren't expected merely to tune in and listen. This network is push and pull, give and take. It broadens our reach. If we lock our keys in the car, the network can unlock it for us from thousands of miles away—just a few bits in the ether.

To play in this game, we must equip ourselves with gadgets. Communicative gadgets: mobile phones, pocket computers, radio-synchronizing wristwatches, remote car keys, smart cards and smart tags, microchips and antennae sewn into our hems and lapels. Never mind the dismal sounds from Wall Street; the share prices of Palm and Nokia are not a leading indicator on this matter. Mobile phones are nearing ubiquity: teenagers depend on them and frenetically instant-message their pals; couples stroll together engaged in parallel telecommunication; the New York–Washington shuttle before departure is one big tubular telephone lounge. But the phone is just the obvious part. IBM is preparing Digital Jewelry: earrings, bracelets, chokers, microphones, cameras, tiny brains, all with tiny batteries, all communicating wirelessly. We're to lodge these items nightly in their Digital Jewelry Box, where they will recharge their spirits and swap data. We children go to sleep; our toys stay up and play.

So the editors have sent me forth, equipped. I have joined what the Japanese are calling the *oyayubizoku*: the thumb tribe, named for the organ we so compulsively poke at our tiny keypads. I am meant to be the Compleat Geek. My hip vibrates with each incoming e-mail, because of the BlackBerry two-way pager clipped there, just like Al Gore's. I have the i-O Display Systems i-glasses, pixels glistening before my eyeballs, one step short of pumping virtual reality directly into my optic nerve. I feel reluctant to wear this item, so sleek and cumbersome, so fashion-forward and yet so retro, out in Times Square. I share my misgivings with the editors. Their reply comes by e-mail: "I'll advise you to look at the fine print in your contract, which specifically addresses the type of headgear we can force you to wear in public."

Oh, fine. I'm connected.

Information everywhere, at light speed, immersing us—is this what we want? We seem unsure. We are the species that defines itself in terms of information: *Homo sapiens sapiens.* We're knowledge connoisseurs. We're being promised some approximation of: All Previous Text (and music and pictures)—in our pockets. Then again, we didn't evolve in a world with so much data and buzz. Our sense organs tuned into one slow channel at a time. Now we tune in and out. The dream of perfect ceaseless information flow can slip so easily into a nightmare of perfect perpetual distraction.

Our technologies don't just empower us; they also harass us, and they change us, for better and for worse. None more than the computer. "Other inventions alter the conditions of human existence," writes Richard Powers in his new novel, *Plowing the Dark*. "The computer alters the human. It's our complement, our partner, our vindication. The goal of all the previous stopgap inventions. It builds us an entirely new home." All the more so when the computer is . . . *everywhere*.

But a long and bumpy path lies between promise and reality. "Wireless" is still a relative term. The cable and plug industry need not panic. Heading out to Silicon Valley with my wireless devices, I find myself assembling the following accessories:

- For my laptop, a three-pin AC adapter plus power cord.
- Cell phone has its own power cord, plus a cable to the laptop. Plus a hands-free headset—earplug and microphone in one—so I can walk along giving fellow New Yorkers the impression that I'm talking to myself. I do not yet have a wearable cellular phone arm band, rapid charger kit, holster, or leather case.
- For my handheld PC, a docking cradle. A power cord and adapter. And a detachable cable for synchronizing the data in the handheld with another PC. The handheld also has a modem attachment, which has its own power cord.
- For my digital-music player, I have plug-type earphones, although my otorhinolaryngologist disapproves of the attempt to apply sound right to the eardrum.

Another cable connects the music player to my laptop, for loading the music in the first place.

- A different (but indistinguishable) cable connects my digital camera to the laptop, for unloading the pictures. My digital voice recorder, too, has a unique cable. No power cord; it runs on AAA batteries.

- Batteries.

- Manuals and warning placards. "To satisfy FCC RF exposure compliance requirements," says one, "the user should generally maintain a separation distance of 4 cm between the person's body, and the device and its antenna." But, relax, we can make an exception for hands, "because they are extremities."

There are old-style connectors: serial ports, with nine pins or twenty-five. There are new-style connectors: USB and FireWire. I try to coil some of the wires. They came with twist ties for this purpose. If I were better organized, I would have a box just for the twist ties. My wife watches dolefully: "You're setting up the Mir space station?"

In the imaginations of the gadget-makers, these cables have already vanished. A new wireless standard called Bluetooth (after Harald Bluetooth, first-millennium Viking king) is meant to replace them. Every gadget gets a Bluetooth chip, with its own radio transceiver. All these Bluetooth-enabled devices sense one another's presence, trade stories, and keep one another up to date. They create spontaneous personal networks, where devices can act simultaneously as master and slave to other devices: ad hoc "scatternets," "personal area networks," networks within networks. Your Bluetooth mobile phone may obey instructions from a concert hall to switch to vibrate-only mode. Your Bluetooth headset will presumably know, at any given moment, whether to keep playing some song you've downloaded or switch to an incoming phone call or alert you to an impending thunderstorm.

Into the same virtual space, the electromagnetic spectrum, comes a wholly different wireless standard, called 802.11b. (Say "eight oh two dot eleven bee.") The proponents of this standard are pushing a friendlier name, Wi-Fi, for wireless fidelity. Apple wires all its new laptops for 802.11b and other manufacturers are following suit. If you install a small base station somewhere on your home network, you can carry laptops from room to room, basement to kitchen counter, and never go off-line. By the end of this year, thousands of hotels, airport lounges, and coffee shops will be filling their airspace with this same invisible radiation field: information and connectivity all around. Microsoft and Starbucks are teaming up to deploy it. One can imagine grocery stores and department stores beaming real-time information to their gadget-toting customers. One can even imagine properly functional motor-vehicles offices and polling places.

Wi-Fi is "the next big thing," asserts J. William Gurley, a Menlo Park, California, venture capitalist and online columnist. "The history of technology has proven again and again that if a certain open architecture gains escape velocity there is no turning back." He's right. Whether he's right specifically about 802.11b doesn't matter. It might be Wi-Fi, or it might be Bluetooth, or it might be a combination of those and something else besides.

Stock-market watchers have their own special perspective, of course, and lately they have been glum. Yet people in the world's silicon valleys are still showing up for work and planning our future. The recent spells of market euphoria and market despair, at their most extreme, have been illusions, with little relation to the real breadth of technological change. We do tend to take our illusions seriously, when they involve large sums of money. For that matter, the rising volatility on Wall Street flows directly from the dense, high-velocity interconnectedness of our information sources. When everyone hears the same "news" at the same time, and everyone tries to buy or sell the same stocks, any hope of market equilibrium vanishes. We are learning to live with whole new species of mass hysteria.

In other ways, too, these developments pose challenges to the life of the polity. More than ever, our ability to participate in the basic processes of our information-rich culture—commerce, education, entertainment—will depend on technology. The Internet has been a democratizing force worldwide, knocking down walls, creating new voices, redistributing knowledge— sometimes, redistributing the kind of knowledge that brings wealth. But there are barriers to entry. Like our other core infrastructures—roads and bridges, the electric power grid, the phone system—the wired and wireless network is being built out largely by private companies, yet the public needs universal access. If laptops and Internet connections and Web-aware mobile phones remain tokens of privilege, then the gap between rich and poor will grow. Digital jewelry, indeed.

The lexicographers of the *Oxford English Dictionary* have an open file on the word *network*. Some of the file is virtual: bits living in the network. Some of it exists in more traditional, detached form, on four-by-six-inch slips of paper, which, at the moment I inquire, happen to be out on someone's desk in Oxford. I'm wondering whether they've tracked this new sense of the word: "the network," or even "the Network," meaning a global entity, bigger than the Internet. The totality of the world's computers, data-bases, and communications channels. Maybe the network can be said to possess knowledge and even behaviors.

Sure enough, they are keeping an eye on *the network*. "It would seem that the new sense you mention is closely—maybe inextri-cably—tied up with a usage which goes back well before the Internet came into existence," says Peter Gilliver, an *OED* associ-ate editor. Not the original sense, of course (*work in which threads, wires, or similar materials are arranged in the fashion of a net*), but something connoting the totality of all information networks—and some-thing we tend to personify ("the network listens"; "the network knows"). Gilliver checks science and science fiction without prej-udice. "As long ago as 1970," he notes, "the network was clearly used in very much your sense—the only difference being that in 1970 that 'totality' was pretty limited."

They might also want to take a new look at the word *pervasive*. I'm hearing it all over Silicon Valley, and without the usual pejo-rative overtones. Pervasive computing is both a buzzword and a new field of study within computer science. It means computers in the walls, in tables and chairs, in your clothing. Computers in the air, when engineers can figure that one out (a group at Berkeley is working on "Smart Dust," financed by the Defense Advanced Research Projects Agency). Computers fading into the environment.

Computer scientists, embracing this vision, see their disci-pline as a new branch of social science. They look back over their shoulders at the humans in the picture, and sometimes they sound surprised: "Individuals within the space are doing things other than interacting with the computer," declares a recent research report, "coming and going, and perhaps most strikingly, interacting with each other—not just with the computer."

Pervasive Computing is also a new division and "strategic initiative" at IBM, spreading across several of the company's headquarters and research laboratories. Helga has guided me past San Jose, around some hills and dales, to the astoundingly bucolic Almaden Research Center, about seven hundred scientists in four hypermodern buildings hidden in a field dotted with cows. I head for the User Lab—the place where they're supposedly thinking about us humans and where we fit in. The head of the User Lab is Daniel M. Russell, a lanky, mischievous-looking man with a trim white beard.

He begins by announcing, "I fundamentally don't care about computers." But he led computer-research groups at Apple and at Xerox, and he has computers on the desk and computers on the wall, and, as his staff wander in and out, they pretty much all have computers in their pockets, and even that skateboard in the corner happens to be a computer on wheels, so clearly there's some kind of subtle distinction coming.

"I care about compu*ting*," Russell says. "I care about what you can do with this thing, this magical property, this thing we've imbued into our devices. This lab is about computing as a medium for people, a medium of expression and a medium for work, and so on."

He's got cognitive psychologists, mechanical engineers, and industrial designers. He's got a working machine shop. All around are bits of gadgeteer detritus: broken-up pagers and wristwatches, eyeglass frames and limbs from department-store manikins.

"One of the pieces of what we're doing is thinking about how can we make devices smaller, and smaller, and smaller," he says. "You can imagine where all this leads, right? The obvious terminal point is you implant them—which brings up its own set of issues." (The jokes about our bionic and cyborg future fly freely around here.) "Or you turn it into jewelry." Left earring talks wirelessly to right earring. Pendants become annunciators, rings become pointing devices or alarms.

"I want my ring to shine red when my daughter gets home," Russell says. "Or flash green when I have an urgent message. Or the stock price shoots above 200. So now the question becomes: once you've got rings that talk to your computer, and cell phones that are in your ears, how do you get them to work together? How do you dial someone?"

Cameron Miner runs the group's "designLab" for working on such questions. "We're seeing a usability gap emerge with these devices," he says—devices constantly shrinking while adding new functions. "My eyes are not getting any better. My fingers are not getting any smaller."

Tiny keyboards are just frustrating. Voice recognition is everyone's dream, but understanding human speech is one of those fundamental capabilities that continue to elude machines. It's a hard problem.

These researchers share certain articles of faith, though. One is that their world marches to the steadfast drumbeat of Moore's Law. However tiny and however powerful this year's devices are, next year's will be tinier and more powerful. They can bank on it. So they're planning ahead. They also believe that no matter how Luddite we feel, deep down we're data addicts, suckers for information. Resistance is futile. With all the stuff being thrown against our walls, some of it has to stick.

It may as well be a law of modern life. Once it was true of machines, as they began infiltrating the fabric of our existence, and now it's particularly true of the technologies of computing and communication. First we disdain them and despise them; then we depend on them. In between, we hardly notice a transition.

Our first wearable information appliance established itself long ago. The wristwatch industry has never been healthier, though some stalwart souls still affect not to wear such a thing. "We find that people look at their watches four dozen times a day," says Miner. "And at no time do you realize that more than when you forget your watch. It's not just that you don't know what time it is; you feel all out of sorts. The rhythm of your day is all thrown off.

"So we're thinking, what other kinds of information can you push into that peripheral channel? Contacts and schedules and things like that are good. But what about your stock-market portfolio? Or biometric information about your loved ones, so you can see how your parents are doing, just to know whether they're having a good day or a bad day."

We wear other devices, too. We've cheerfully sported lenses in front of our eyes for several hundred years. They could be smarter. Russell has prototype eyeglasses that translate signs from Japanese into English, displaying the translation as a caption, a half-inch from your retina. Now, translation is another of those hard artificial-intelligence problems, but still. "Even if the translation is terrible," Russell says, "I don't read any Japanese at all, so for me, *this* is a lot better than *that*."

Pervasive computing isn't just about gadgets to carry and wear, though. These researchers are thinking about our whole environment. They have rooms that use tiny cameras to watch people's eyes and keep track of what we're looking at. They're conducting studies of how we behave, and how we feel about it, when we can glance at an appliance and say, "Turn it on." They assume that entire houses will be ensembles of hidden computers.

The head of IBM Pervasive Computing is Michel Mayer, a product of the École Supérieure d'Électricité in Paris and a company veteran. "It's going to be more and more machines talking to machines, things talking to things, without human interaction," he says. "We're already there. The infrastructure, although it's boring and more remote and in the background, is increasingly important. It's going to be your fridge, your car, your tools, your clothes, doing all those little microelements of tasks. It's going to be your dishwasher negotiating with your utility company over what the best rate is and when."

The average American house already contains more than forty computers embedded in various items. A typical electric toothbrush runs on about 3,000 lines of code. Last year alone, 8 billion new microprocessors came into the world. "These are mostly brain-dead right now," Russell says, "they're tiny, four-bit processors, and so on. But you *know* where our world is going."

Even here, in this bastion of cheery futurism, they don't assume this is unalloyed good news. Science fiction writers have been warning for years about this sort of world: painting scary pictures of a human race dependent on technology that runs amok, or just breaks down.

"Well, yeah!" says Russell. "We read that stuff, too! How many times has your cell phone crashed on you? Mine crashed last night. When your house crashes, how do you restart your house?" This question doesn't have a good answer, although, with smaller gadgets, we've learned to find the reset button or yank the batteries to cycle the power. And pray it won't be necessary to phone for technical support.

With new possibilities come new anxieties. How much smarter do you want your house to be, when you still haven't mastered setting the time on your VCR, your stove, and your coffeemaker? Then, if the devices learn to reset their own clocks remotely, will you trust them? One central modern fear is that as machines grow too complex to understand and repair easily, we grow helplessly reliant on them. We become their slaves. This was the main argument of Ted Kaczynski, the Unabomber, but that doesn't mean it's completely insane.

It's certainly time to worry about privacy and personal autonomy. If your truck is GPS-equipped, or your car has an electronic toll-paying tag, the network is already capable of keeping track of your whereabouts, so you may not care to implant a tracking chip under your skin. But you could. Your employer may already be testing electronic tags and badges for this purpose.

And every new channel of information is a potential intruder with a sales pitch. Maybe we've gotten used to advertisements next to magazine articles. Maybe we can even handle billboards in public airspace, and commercials at movie theaters where we've paid for seats, and telephone promotion from companies keeping us on hold. It's going to get worse, quickly. You will soon notice lots of little screens beaming messages at you. On airplane seatbacks: *"Improved Data Speed!!! . . . Turn your e-mail into voice mail . . . Nasdaq -31.38 . . . Real-Time Stock Quotes . . . Was weather something you planned for? Select Weather Channel . . ."*

"When displays become essentially free, they're all going to be subject to sale," says Russell. "I have a little display in my home thermostat. Believe me, the thermostat company's going to want to put in animated graphics, and if they can possibly sell that space to the heating-oil company, they'll do it. So one of the questions about ubiquitous technology becomes: Who owns your attention? Who owns the right to *push* inside your personal environment? When you walk past a store, your cell phone could say, *Come in! Ten percent off!* How do you screen that stuff? How do you anti-spam-filter your life?"

The pervasive-computing people are breaking the computer apart. Every function—speak, locate, photograph, read, remember—can be detached. They think of it as the constellation model of computing. Your devices form a constellation. They all talk together, and they don't need much transmission power because they only have to cover the distance from ear to pocket, say. Displays, processors, memory, power can all be separated. It's efficient. We already carry amazing amounts of spare processing power, in our cell phones if nowhere else.

Turbulent crosscurrents here. We see the computer splitting into its constituent parts, which can float more or less freely. At the same time we see all the different components combining and recombining. Combo digital camera and digital music players are hitting the market. A cell phone available in Hong Kong doubles as an ovulation clock and calorie counter. A Global Positioning System chip can meld with almost anything. The building blocks of electronic life are suddenly . . . *building blocks,* and manufacturers want to try one permutation after another. We consumers, meanwhile, exhibit signs of craving the single perfect gadget—the Swiss Army Knife of digital devices. So which will it be? Free-floating specialization, or all-functions-in-one?

This is one of Jeff Hawkins's favorite problems. If any one person can be said to have invented the Palm Pilot—the handheld computer that defined the entire product category—it is Hawkins, a loose-limbed, perennially boyish electrical engineer who carved his first prototype from a block of wood and carried it around pretending to scribble on it. He and Donna Dubinsky founded Palm Computing in 1992, sold a million Pilots in a year and a half, sold the company to US Robotics, and left in 1998 to found another Silicon Valley start-up, called Handspring. Handspring makes handheld computers much like Palm Pilots, called Visors. They have instantly grabbed 25 percent of the domestic market.

"There's a yin and yang going on here," says Hawkins. "I think people would like to have one thing. Sometimes people come to me pleading! *Please can you make it one thing.* The other side of that is that today maybe we can't make a good One Thing. Maybe when you try to combine them you end up with not-so-good products. The best voice recorder"—he's gesturing at my digital recorder, which is barely bigger than a cigarette lighter—"might be compromised if you try to make it into a cell phone *and* a handheld computer."

Nonetheless Handspring has started shipping its combo cell phone and handheld computer, the VisorPhone. The telephone is actually a module that pops in and out of the Visor's expansion slot, so when you're finished with a call, you can pop in a different module to turn the gadget into a camera or satellite navigator.

"It's all temporary, in my mind," says Hawkins, "but in the mobile device space, my bet is on the singular device. You know about Bluetooth? Everyone says all these little things will talk to one another and wow it'll be great. I don't buy that. It's too complex. I'm not so much into bodywear and smart clothing, and so on. No. You can carry something. I don't think we're going to see the retinal implant or the 3D glasses or whatever. We will have *beautiful* color displays. Of course it's all going to be wireless."

We should begin taking it for granted that we'll all have high-bandwidth connections in our pockets. "What's really interesting, and I don't think most people have understood," he says, "is that it will be free."

Maybe I look dubious. Hawkins continues: "People say, 'Well, how can it be *free*?' All right, it won't be *completely* free, but it will be free like local telephone service is free: yeah, I pay my eight bucks a month, but I don't think twice when I make a call.

"It's costing billions of dollars to build it all out, but the incremental cost of adding a customer is very, very small. It's all virtual, it's just more bits going one way versus the other way. The cost for both voice and data will be virtually free, in the whole wireless connected world."

Information is convenience, and information is power—that's a given, around here. "There's a fundamental premise here that communications and computing technology is a net benefit for people," Hawkins says. "Not in all cases, but for the vast majority."

This means that the most fundamental social processes are poised for transformation. Voice communication, taken as a whole, is "crude." Voice mail is essentially "broken." Money is up for grabs.

"The whole concept of writing checks!" Hawkins says. "This is bizarre!" He's excited now. Really, the old unwired world was so baroque. "You know, I have this book of *papers,* and I've got to *order* them, and I've got to write them in and rip them out, I've got to log them, I have to take out my calculator, month to month they send them back to me, people *handling* them all along the way! This is a system that's ripe to be replaced!

"Exactly how it gets replaced, I don't know," he says. "But I would argue that it will be *a mobile device that's wirelessly connected.*"

Is this getting scary, at all? How much of the planet do I want connected to my checkbook at any given moment?

"Of course it has to be secure," Hawkins says. With a flourish, he yanks a wallet from his back pocket. "*This* thing isn't very secure, either. People say, 'Oh, I would never carry around a device that has, you know, access to my personal information—what if I lose it?' "

He puts his wallet back. "I think they'll get over that."

Somewhere in the same family tree, along with the handheld computers and the mobile phones, sits a seemingly different gadget, the remote control. We take our remote controls for granted these days, but they continue to creep deeper into our lives. It can feel disconcerting—oddly passive or powerless—to watch television without a remote control in hand. Some new homes come with remote controls for dimming the lights. They lengthen our arms. The modern car key is a remote control and, for that matter, a wireless computer. One need not actually touch the car to unlock it, start the climate control, or reset the alarm. We seem to grow fond of waving a wand, or thumbing one, and seeing things happen at a distance.

My own car key is so complex—three buttons, which can be pressed or pressed-and-held—that I need to keep the instructions close by. In theory, at least, I can open and close windows. I can make the seat and mirrors return to a specific position, all from across the street. This saves me the trouble of pressing a button on the seat. To get it all set up, though, required a long visit to the dealer's service department. The car spent an hour linked to a computer, getting some sort of data transfusion. I spent the hour in the waiting room, thinking about the time savings that would accrue over the years to come, one millisecond after another. I couldn't stop wondering whether we're like some species of deer that's going to evolve bigger and bigger antlers, until we're the best antlered animals around, but we can no longer lift our heads out of the mud, so we go extinct.

Most wireless gadgetry isn't quite ready for mass consumption. Most of it works only sporadically and only in certain places. All of it comes with hidden costs not listed on the boxes: time the consumer must invest in reading manuals, managing batteries, coiling the supposedly nonexistent wires, and generally learning new skills. At its best, browsing the Internet on a Web-enabled phone feels like looking through the wrong end of a telescope. No wonder some people assert almost religiously that they will *never* use a cell phone, or a handheld computer, or a stereoscopic 3-D optical headset with optional immersion visor.

But five years ago, some of the same people felt no need for e-mail or call waiting. The clumsiness and inelegance will pass, as with all new technologies. Some people will adapt after all. Some of us will just die off and be replaced by the next cohort, young enough not to remember a world without e-mail. Hawkins laughs when I say this: "I was kinder—I said 'generational changes.' Yeah, I'm amazed at how quickly young people adapt."

It's an industry truism that children most readily learn the necessary new styles and habits. They know to power-cycle gadgets that crash, and they instantly acquire the most esoteric special typing skills. "They're mutants! They're cyborgs!" says Michel Mayer at IBM. "I don't know how they do it!" He doesn't seem all that unhappy about it.

Another lesson of the television remote control is that no one can predict how we're going to use new technologies, much less tell us how to use them. The inventors of the remote control believed consumers would use them several times an evening—to turn the set on, to change the channel when a program ended, to turn the set off. No one imagined that we would become remote-control virtuosi, personal entertainment maestros, creators of our own nightly medleys. Even when we feel deluged and assailed by technology—even when we suspect marketers of foisting useless gizmos upon us—we tend to make our own choices. Perhaps handheld gadgets will offer the kind of power over the rest of our experience that the remote control gives us over TV: the power to edit and jump—instant access, fluid montage, snippets and shards. For better and for worse.

Technologists tend to believe that we're actually smarter for having these gadgets and that, as they permeate the texture of modern life, we will grow smarter still. That's a collective, grand, slightly murky *we*. Bernardo Huberman, scientific director of the Sand Hill Labs—a new Hewlett-Packard research center—talks about harnessing social knowledge: "studying the whole Internet ecosystem and designing novel mechanisms and institutions so that we can harvest the distributed knowledge that such a gigantic social mind is producing."

We don't have to become neurons in the new world brain to feel that we're already gaining something. I've noticed that the mobile gadget-wielder develops the odd sensation of being entitled to all sorts of facts. You get in the habit of knowing things, or at least of being able to find out. It's as if there's a permanent mental hotline to the information specialists at the public library. Can't quite identify Bob Dole's running mate in 1996, or that actor up on the screen, or a science fiction story encountered ten years ago? You get a twitchy feeling that you ought to push a button and pop up the answer.

But Huberman has more in mind than facts and trivia. His research consistently finds informal communities making better decisions than any of their members—knowing more and thinking better than experts. "We now know that society can work better than any individual," he says. "There is this notion of a collective mind, a social mind, and today the Internet allows us to tap that." We are distributing intelligence. We are creating social organisms that carry out continuous computation.

It may be true, even if we don't see it moment by moment. So I drive along, giving Helga another chance, wondering whether to check the e-mail that's even now tingling at my hip. If I can manage this, with one hand, I'll see a message from **the commander and head of the Secret Unit in charge of Diamond dealing for the Revolutionary United Front (RUF) of Sierra-Leone.** It seems the commander would like my help in disposing of a **large quantity of diamond and about US\$12,500,000.00 that is in cash with my Wife who does not have the know-how to launder this money.** I need only send along my bank information. Memo to self: add commander to anti-spam filter.

It would be fair to wonder, meanwhile, how well I'm focused on driving the car. I've learned that people with dashboard navigation systems often keep an eye on the moving map even when they know perfectly well where they're going. Why? For entertainment? To keep from being . . . bored? We work hard to avoid being alone in our heads. Yet things happen when we're alone in our heads and they're things we can't live without: contemplation, reflection, focus. As consumers of wireless gadgets, we will need to insist on a Disconnect button. My BlackBerry has one, mainly for use on airplanes. Helga must have one, but I haven't found it yet.

After a while I also catch myself wondering whether—if I had to—I could still unfold a paper map and find my own way. Satellite navigation really works, and one could easily learn to depend on it. "Yeah, you can regret that you don't know how to read maps anymore," says Mayer. "But all in all I think it is progress somehow. Mankind is hungry for new capabilities."

So we engage in this ritual—the never-ending, reluctant dance with the invader. Technology encroaches and we resist. Our aversion is sensible and honorable. And then, later, we give up. In this case, we connect.

Acknowledgments

In the matter of editors, I have been lucky. I am indebted and deeply grateful to Jack Rosenthal, Katherine Bouton, and Adam Moss; David Remnick and Jeffrey Frank; and Dan Frank, as well as my agent, Michael Carlisle.

About the Author

James Gleick is the author of *Chaos, Genius,* and *Faster.* A native of New York City, he graduated from Harvard College in 1976 and worked for the *New York Times* for ten years as an editor and reporter. He was founder and CEO of the Pipeline, one of the first Internet access services. His writing on science and technology has been translated into more than twenty-five languages. He lives with his wife in the Hudson Valley of New York.